Teaching Statistics Using Baseball

© 2003 by
The Mathematical Association of America (Incorporated)
Library of Congress Catalog Card Number 2003103273

ISBN 0-88385-727-8

Printed in the United States of America

Current Printing (last digit):
10 9 8 7 6 5 4 3 2 1

Teaching Statistics Using Baseball

Jim Albert

Published and Distributed by
THE MATHEMATICAL ASSOCIATION OF AMERICA

Classroom Resource Materials is intended to provide supplementary classroom material for students—laboratory exercises, projects, historical information, textbooks with unusual approaches for presenting mathematical ideas, career information, etc.

MAA Service Center
P. O. Box 91112
Washington, DC 20090-1112
1-800-331-1622   fax: 1-301-206-9789

Contents

Preface

Over twenty years, this author has been in the enterprise of teaching introductory statistics to an audience that is taking the class to satisfy their mathematics requirement. This is a challenging endeavor because the students have little prior knowledge about the discipline of statistics and many of them are anxious about mathematics and computation. Statistical concepts and examples are usually presented in a particular context. However, one obstacle in teaching this introductory class is that we often describe the statistical concepts in a context, such as medicine, law, or agriculture that is completely foreign to the undergraduate student. The student has a much better chance of understanding concepts in probability and statistics if they are described in a *familiar* context.

Many students are familiar with sports either as a participant or a spectator. They know of the popular athletes, such as Tiger Woods and Barry Bonds, and they are generally knowledgeable with the rules of the major sports, such as baseball, football, and basketball. For many students sports is a familiar context in which an instructor can describe statistical thinking.

The goal of this book is to provide a collection of examples and exercises applying probability and statistics to the sport of baseball. Why baseball instead of other sports?

- Baseball is the "great American game." Baseball is great in that it has a rich history of teams and players, and many people are familiar with the basic rules of the game. The popularity of baseball is reflected by the large number of movies that have been produced about baseball teams and players.

- Baseball is the most statistical of all sports. Hitters and pitchers are identified by their corresponding hitting and pitching statistics. For example, Babe Ruth is forever identified by the statistic 60, which was the number of home runs hit in his 1927 season. Bob Gibson is famous for his record low earned run average of 1.12 during the 1968 season.

A flood of different statistical measures are used to rate players and salaries of players are determined in part by these statistics. There is an active effort among baseball writers to learn more about baseball issues by using statistics.

- A wealth of baseball data is currently available over the Internet. Player and team hitting and pitching statistics can be easily found. Comparisons between players of different eras can be made using a downloadable dataset that gives hitting and pitching data for all players who have ever played professional baseball.

This book is organized using the same basic organization structure presented in most introductory statistics texts. After an introductory chapter, there is a chapter on the analysis on a single batch of data, followed by chapters on the comparison of batches, and the analysis of relationships. There are chapters on introductory and more advanced topics in probability, followed by topics in statistical inference. Each chapter contains a number of "essays" or "case studies" that describe the analysis of statistical or probabilistic methods to particular baseball data sets. After the collection of case studies in each chapter, there is a set of activities and exercises that suggest further exploration of baseball datasets similar to the analysis presented in the case studies.

How can this book be used in teaching probability or statistics? We suggest several uses of this material.

- This book can be used as the framework for a one-semester introductory statistics class that is focused on baseball. Such a class has been taught at the author's home institution. This course covers the basic topics of a beginning statistics course (data analysis, introductory probability, and concepts of inference) using baseball as the primary source of applications. This course is suitable for students who are interested or curious about the game of baseball. It is also suitable for students with sports-related majors, such as sports management or sports medicine.

- This book can also be used as a resource for instructors who wish to infuse their present course in probability or statistics with applications from baseball. The material in this book has been presented at different levels to make it useable for introductory and more advanced courses. The case studies can be used by the instructor to present the particular topic within a baseball context and then the associated exercises and activities can be used for homework. The case studies can serve as useful springboards for undergraduate students who wish to do additional explorations on baseball data.

Acknowledgments

I am appreciative of the support given to this project by the Division of Undergraduate Education of the National Science Foundation and by my colleagues in the Department of Mathematics and Statistics at Bowling Green State University. The text was used for a number of experimental sections of MATH 115 *Introduction to Statistics* at Bowling Green State University and I am grateful for the valuable feedback from the students who enrolled in this course. In addition, Chris Andrews, Jay Bennett, Eric Bradlow, Jim Cochran, Joe

Gallian, Carl Morris, Jerome Reiter, Ken Ross, Steve Samuels, Bob Wardrop, and Dex Whittinghill provided many helpful suggestions in reviewing the book. I thank the editors of the MAA publications for their support, particularly Zaven Karien and Dave Kullman. Last, but certainly not least, I thank my wife Anne, and children Lynne, Bethany, and Steven for their understanding and great patience during the completion of this text.

JIM ALBERT
December 2002

1

An Introduction to Baseball Statistics

Leading Off

The baseball game has just started. The umpire has yelled "Play Ball!" and the batter at the top of the order is coming to bat. He's the leadoff hitter and his job is to help produce runs by getting on base. Who was the greatest leadoff hitter of all time? Most people believe that the best leadoff man was Rickey Henderson. Bill James, a leading baseball statistician, says in *The New Bill James Historical Baseball Abstract* that Rickey was

- the greatest base stealer of all time,
- the greatest power/speed combination of all time (except maybe Barry Bonds),
- the greatest leadoff man of all time,
- one of the top five players of all time in runs scored.

Moreover, James says

> You could find fifty Hall of Famers who, all taken together, don't own as many records, and as many important records, as Rickey Henderson.

Here's some biographical information about Rickey Henderson. He was born on Christmas Day, 1958, in Chicago, one of seven children. His family moved to Oakland, California when he was young and Rickey played baseball and football at Oakland Tech High School. When he graduated, he received many football scholarships and also was selected by the Oakland A's in the fourth round of the 1976 baseball draft. Although Rickey preferred football, his mother wanted him to play baseball, and Rickey agreed to go along with his mother's wishes. After a couple of years in the A's minor league organization, he was promoted to the Oakland major league team on June 23, 1979[1] and immediately was a starter

---

[1] I have a kinship with Rickey Henderson since we both started professionally (me as a statistician and Rickey as a ballplayer) in 1979 and we're both still active in our jobs. I suppose that I will last longer than Rickey in the professional ranks.

on the team. He has been a dominant player in the major leagues his entire career and was voted on the All-Star Team for the years 1980, 1982, 1983, 1984, 1985, 1986, 1987, 1988, 1990, and 1991. How can we demonstrate that Rickey Henderson was indeed the best lead-off man in baseball? We look at his stats. Table 1.1 displays the season-to-season hitting statistics for Rickey for his first 23 seasons in Major League Baseball.

TABLE 1.1

Batting statistics for Rickey Henderson for the first 23 seasons of his career.

Season	TM	G	AB	R	H	2B	3B	HR	RBI	BB	SO	SB	CS	AVG	OBP	SLG	OPS
1979	Oak	89	351	49	96	13	3	1	26	34	39	33	11	.274	.338	.336	.674
1980	Oak	158	591	111	179	22	4	9	53	117	54	100	26	.303	.420	.399	.819
1981	Oak	108	423	89	135	18	7	6	35	64	68	56	22	.319	.408	.437	.845
1982	Oak	149	536	119	143	24	4	10	51	116	94	130	42	.267	.398	.382	.780
1983	Oak	145	513	105	150	25	7	9	48	103	80	108	19	.292	.414	.421	.835
1984	Oak	142	502	113	147	27	4	16	58	86	81	66	18	.293	.399	.458	.857
1985	NYY	143	547	146	172	28	5	24	72	99	65	80	10	.314	.419	.516	.935
1986	NYY	153	608	130	160	31	5	28	74	89	81	87	18	.263	.358	.469	.827
1987	NYY	95	358	78	104	17	3	17	37	80	52	41	8	.291	.423	.497	.920
1988	NYY	140	554	118	169	30	2	6	50	82	54	93	13	.305	.394	.399	.793
1989	NYY/ Oak	150	541	113	148	26	3	12	57	126	68	77	14	.274	.411	.399	.810
1990	Oak	136	489	119	159	33	3	28	61	97	60	65	10	.325	.439	.577	1.016
1991	Oak	134	470	105	126	17	1	18	57	98	73	58	18	.268	.400	.423	.823
1992	Oak	117	396	77	112	18	3	15	46	95	56	48	11	.283	.426	.457	.883
1993	Oak/ Tor	134	481	114	139	22	2	21	59	120	65	53	8	.289	.432	.474	.906
1994	Oak	87	296	66	77	13	0	6	20	72	45	22	7	.260	.411	.365	.776
1995	Oak	112	407	67	122	31	1	9	54	72	66	32	10	.300	.407	.447	.854
1996	SD	148	465	110	112	17	2	9	29	125	90	37	15	.241	.410	.344	.754
1997	Ana/ SD	120	403	84	100	14	0	8	34	97	85	45	8	.248	.400	.342	.742
1998	Oak	152	542	101	128	16	1	14	57	118	114	66	13	.236	.376	.347	.723
1999	NYM	121	438	89	138	30	0	12	42	82	82	37	14	.315	.423	.466	.889
2000	Sea/ NYM	123	420	75	98	14	2	4	32	88	75	36	11	.233	.368	.305	.673
2001	SD	123	379	70	86	17	3	8	42	81	84	25	7	.227	.366	.351	.717
Total	—	2979	10710	2248	3000	503	65	290	1094	2141	1631	1395	333	.280	.402	.420	.822

The stats that most people talk about are Rickey's career totals.

- He stole the most bases (1395) of any baseball player in history.

- He scored the most runs (2248) of any player in history.

- He received the most walks (2141) of any player in history.

Those are great achievements and the statistics are a measure of these achievements. But the goal of this book is to look deeper at baseball statistics.

There is a general confusion about the meaning of "statistics." If you look at the *American Heritage Dictionary of the English Language*, you'll find two very different definitions of statistics.

1. *Statistics* is a collection of numerical data.
2. *Statistics* is the mathematics of the collection, organization, and interpretation of numerical data.

Let's relate these two definitions to baseball. First, baseball statistics are the counts and measures that we use to evaluate players and teams—this refers to the first definition of statistics. When the announcer on a television broadcast of a baseball game thanks the statistician, he or she is referring to the guy who collects and gives baseball data to the announcers. Here the focus is on the data.

But this book concentrates on the second definition of statistics—how can we interpret or make sense of baseball stats? A professional statistician (to be distinguished from the guy who is collecting the data) is interested in how we can use data to learn about some underlying truth. In doing this, he or she has to think about several issues.

- How should the data be collected to make it useful in drawing conclusions?

- Once the data is collected, how do we organize and summarize it to learn about its general features?

- Last, how can we use the data to make our conclusions? (It turns out that probability or chance plays an important role in decision-making.)

It might be helpful to distinguish the two meanings by capitalization—in this section I will call numerical data *statistics,* and the science of learning from data *Statistics.*

The goal of this book is to introduce Statistical thinking and Statistical methods in the context of baseball. Let's introduce the chapters of this book by looking at the statistics of Rickey Henderson. Some questions will be raised in the following discussion and we'll continue our Statistical look at Rickey Henderson by "leadoff exercises" in each chapter.

Exploring a Single Batch of Baseball Data (Chapter 2)

The goal of a leadoff hitter is to get on base. The obvious measure of a player's ability to get on base is the on-base percentage (OBP), which is simply the fraction of plate appearances where the player gets on base. (A precise definition of OBP will be given later.) Here are Rickey's season OBPs for his first 23 seasons in the majors:

0.338	0.420	0.408	0.398	0.414	0.399	0.419	0.358	0.423
0.394	0.411	0.439	0.400	0.426	0.432	0.411	0.407	0.410
0.400	0.376	0.423	0.368	0.366				

If we scan these numbers, we see variation—one season he had an OBP of .432 and another season his OBP was .366. A Statistician will try to make sense of these data by constructing an appropriate graph. Figure 1.1 shows a dotplot of the OBPs.

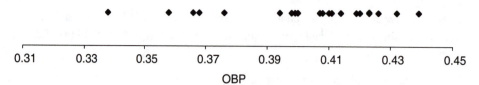

FIGURE 1.1

Dotplot of season on-base percentages for Rickey Henderson between 1979 and 2001.

We see from this graph that most of Rickey's season OBPs are between .400 and .420. There is a small cluster of values in the .340 to .380 range. Why? Was Rickey hurt these particular seasons? Did these low OBPs correspond to the early or late periods of his career? (We'll answer these questions later.)

After we graph the data, we try to find suitable numbers to summarize the main features of the distribution of OBPs. Looking at the graph, it seems that .410 might be a representative season OBP for Rickey.

Comparing Batches and Standardization (Chapter 3)

We now have some handle on Rickey's on-base ability—he generally got on base about 40% of the time. But is an on-base percentage of .400 any good? How does his on-base performance compare with other players?

Another good leadoff hitter, a contemporary of Rickey, was Tim Raines. How effective was Raines in getting on-base and how did he compare with Rickey?

Here are Raines' on-base percentages for the 1981–1998 seasons where he had at least 200 at-bats.

0.391	0.353	0.393	0.393	0.405	0.413	0.429	0.350	0.395
0.379	0.359	0.380	0.401	0.365	0.374	0.383	0.403	0.395

How does this batch of OBPs compare to the batch of OBPs for Rickey? In Chapter 3, we'll compare two batches of data. We will discuss methods for graphing the two datasets and ways of stating Statistically (in this example) that Rickey is better in getting on-base than Raines. This chapter will also show how we can judge the greatness of Rickey's on-base performance in the context of all players that particular year.

Relationships between Measurement Variables (Chapter 4)

When we look at Rickey Henderson's hitting statistics in Table 1.1, we notice that many of the statistics are related. For example, if a batter gets many doubles and triples, he will have a high slugging percentage, and a batter who rarely walks is likely to have a small on-base percentage. In Chapter 4, we discuss ways of looking at relationships between variables. To illustrate, consider the relationship between Rickey's count of doubles and his count of home runs for individual seasons. Doubles and home runs go hand in hand—one might think that if Rickey is hitting a lot of deep fly balls one season then he would have many doubles and home runs. We can graphically view the relationship between doubles and home runs by a scatterplot shown in Figure 1.2. There is a pattern in the scatterplot as the points drift from the lower left to the upper right sections—in seasons where Rickey hit many doubles, he tended also to hit many home runs.

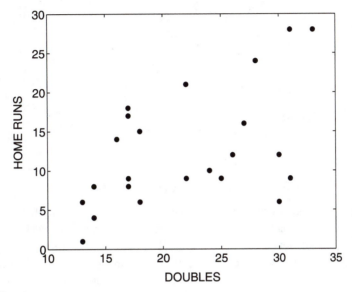

FIGURE 1.2
Scatterplot of count of doubles and count of home runs for the first 23 seasons of Rickey Henderson's career.

In Chapter 4, we discuss ways of measuring the pattern of association in the scatterplot, and discuss how we can use a line to describe the relationship. The methods are helpful for finding a good measure of the batting performance of a player.

Introduction to Probability Using Tabletop Games (Chapter 5)

Fans love to play baseball games—currently millions of people are playing fantasy and simulation baseball. Before there were personal computers, Nintendo, and fantasy baseball, there were a number of tabletop baseball games that were very popular among fans. In Chapter 5, we introduce the notion of a probability model by examining several tabletop baseball games. One of the first games played by the author as a child was *All Star Baseball*, where the performance of a batter was represented by a random spinner with areas of the spinner corresponding to the different outcomes of a plate appearance. A Rickey Henderson spinner is shown in Figure 1.3 where the areas of the regions are computed using his batting statistics from the 1990 season. Note the large pie slice corresponding to a walk—this is a visual demonstration of Rickey's ability to draw walks this particular season.

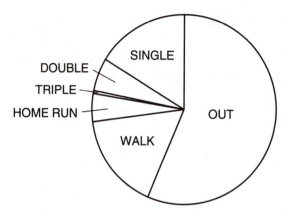

FIGURE 1.3
A Rickey Henderson spinner using his 1990 hitting statistics.

Probability Distributions and Baseball (Chapter 6)

Baseball has a very nice discrete structure that makes for convenient probability modeling. The basic event in baseball is the outcome of a plate appearance. Rickey Henderson comes to bat—either he will get on base or he won't. If you assume that Rickey has a probability of, say .4, of getting on base for each plate appearance, and moreover, the chance of getting on base does not depend on how he hit in earlier games, then one can make reasonable predictions about the number of times he'll get on base in a game, a week, or a month. In Chapter 6, we show how some popular probability distributions (such as the binomial and negative binomial) can be used to explain the random process of players getting on-base, and then scoring runs.

Introduction to Statistical Inference (Chapter 7)

Rickey Henderson had a great *ability* to get on-base. What does this mean? To be an effective leadoff hitter, Rickey had to make the bat hit the ball. (Such a player is called a good "contact hitter.") Also, he knew how to "work the count"—he would not swing at many bad pitches and would patiently wait for a good pitch to be thrown.

How do we know that Rickey had great leadoff hitter ability? We looked at his on-base percentages over his career and saw a pattern of high numbers. In other words, Rickey exhibited a high level of *performance* over many years, and there was no doubt that his performance reflected great skill. It was very unlikely that he had average ability to get on base and, by luck or chance variation, he happened to get high on-base percentages for all of those years.

In Chapter 7, we look at the connection between a player's hitting ability and the performance of the hitter over a season. Actually, if a player has an on-base percentage of OBP = .400 for only a single season, we really don't know if the player has a great ability to get on base. We need to see a pattern of great hitting performance over many seasons to properly gauge the player's hitting ability.

Topics in Statistical Inference (Chapter 8)

Making judgments about a player's hitting or pitching ability is relatively easy when you observe the player's performance over a 10–20 year career. But there are other aspects of ability that are much harder to detect. To illustrate, consider the following *situational statistics* for Rickey for the 1999 season. This season, Rickey's on-base percentage was .423. However, we learn that

- his OBP was .376 for home games and .462 for away games,

- his OBP was .490 for games played on artificial turf and .408 for games played on natural grass,

- his OBP was .472 for games played in a domed ballpark and .417 in open ballparks,

- his OBP was .509 when the pitch count reached 3-2, and .338 when the pitch count reached 1-2.

How do we make sense of these situational hitting stats? Is Rickey really better in getting on base when he is playing away? Does Rickey really have better on-base ability when the game is played on turf? Is it meaningful that his on-base percentage is over .500 when the pitch count gets to 3 balls and 2 strikes? Does Rickey really have an ability to perform extra well in particular situations?

One main topic of Chapter 8 is to try to interpret the significance of situational hitting data. We'll see that it is difficult to detect situational hitting ability and much of the variation in situational hitting statistics is attributable to chance or random variation.

Modeling Baseball Using a Markov Chain (Chapter 9)

We say that Rickey Henderson is the greatest leadoff hitter of all time, but we haven't talked about how well Rickey performed when he was the leadoff hitter in an inning. When we read from Table 1.1 that Rickey had an OBP of .439 in the 1990 season, this tells us the fraction of plate appearances that Rickey got on base *for all situations*. What was his on-base percentage in the innings when he was leading off?

Fortunately, this type of data is now readily available. Let's focus on the home games in Oakland in 1990. In these games, Rickey had 273 plate appearances, but only 108 of them occurred at the beginning of an inning. So Rickey only led off 108 times. How did he do in these lead-off opportunities? Table 1.2 shows the batting results. Note that he was out 63% of the time, so his on-base percentage when he actually was leading off the inning was .370. This on-base percentage appears to be a bit low, but we have to be cautious about drawing a strong conclusion since this table summarizes the results of only 108 plate appearances.

We can also see how Rickey performed when the bases were empty with one out, or when the bases were loaded with two outs. We will use data such as this in Chapter 8 to construct a sophisticated probability model for the sequence of batting events in baseball. This model will be very useful for measuring the values of different types of hits such as a home run and for evaluating the worth of different baseball strategies such as a sacrifice bunt.

TABLE 1.2
Batting results for Rickey Henderson's leadoff plate appearances at home during the 1990 season.

Play	Count	Percentage
Out	68	63.0%
Walk	18	16.7%
Single	13	12.0%
Double	5	4.6%
Triple	1	0.9%
Home run	3	2.8%

Some Basic Measures of Baseball Performance

The effectiveness of batters and pitchers is typically assessed by particular numerical measures. Here we define some basic measures for evaluating hitters and pitchers.

Measures for Batters. The classical measure of hitting effectiveness for a player is the batting average (AVG) that is computed by dividing the number of hits (H) by the number of at-bats (AB):

$$AVG = \frac{H}{AB}.$$

This statistic gives the proportion of time that a batter gets a hit among all at-bats. The batter with the highest batting average during a baseball season is called the batting champion that year. Batters are also evaluated on their ability to get singles (1B), doubles (2B), triples (3B), and home runs (HR). The slugging average (SLG) is an average of a player's "total bases" (TB) where the hits are weighted by means of the number of bases reached:

$$SLG = \frac{1 \times 1B + 2 \times 2B + 3 \times 3B + 4 \times HR}{AB} = \frac{TB}{AB}.$$

This measure reflects the ability of a batter to hit the ball a long distance. A third measure of hitting ability is the on-base percentage (OBP), which is defined as the proportion of plate appearances where the player gets on base.

$$OBP = \frac{H + BB + HBP}{AB + BB + HBP + SF}.$$

In this formula, BB is the count of walks, HBP is the number of times the batter was hit by a pitch, and SF is the number of sacrifice flies.

Measures for Pitchers. A number of statistics are also used in the evaluation of pitchers. For a particular pitcher, one counts the number of games in which he was declared the winner (W) or loser (L) and the number of runs allowed. Pitchers are usually rated by means of the earned run average (ERA)—the average number of "earned" runs (ER) allowed for a nine-inning game:

$$\text{ERA} = 9 \times \frac{\text{ER}}{\text{IP}}.$$

(In this formula, IP is the number of innings pitched.) Other statistics are useful in understanding pitching ability. A pitcher who can throw the ball very fast (such as Randy Johnson) can record a high number of strikeouts (SO). A pitcher who is "wild" or relatively inaccurate in his pitching will record a large number of walks (BB).

Better Measures of Hitting Ability

There is lively research by people to better interpret baseball statistics. *Sabermetrics* is the mathematical and statistical analysis of baseball records. (The term sabermetrics was first used by Bill James in honor of SABR, the Society of Baseball Research.) One interest of sabermetricians (the people who analyze baseball statistics) is to find good measures of hitting and pitching performance. Bill James compares in his *1982 Baseball Abstract* the batting records of two players, Johnny Pesky, who played from 1942 to 1954, and Dick Stuart, who played in the 1960s. Pesky was a batter who hit for a high batting average but hit few home runs. Stuart, in contrast, had a modest batting average, but hit a high number of home runs. Who was the more valuable hitter? James argues that a hitter should be evaluated by his ability to create runs for his team. From an empirical study of a large collection of team hitting data, he established the following *runs created* (RC) formula for predicting the number of runs scored in a season based on the number of hits, walks, at-bats, and total bases recorded in a season:

$$\text{RC} = \frac{(\text{H} + \text{BB}) \times \text{TB}}{\text{AB} + \text{BB}}.$$

This formula reflects two important aspects in scoring runs in baseball. The count of a team's hits and walks reflects the team's ability to get runners on base. The count of a team's total bases reflects the team's ability to move runners that are already on base. This runs created formula can be used at an individual level to compute the number of runs that a player creates for his team. In 1942, Johnny Pesky had 620 at-bats, 205 hits, 42 walks, and 258 total bases; according to the formula, he created 96 runs for his team. Dick Stuart in 1961 had 532 at-bats with 160 hits, 34 walks, and 309 total bases for 106 runs created. The conclusion is that Stuart in 1961 was a slightly better hitter than Pesky in 1942 since he created a few more runs for his team.

There are a number of alternative measures that have been proposed to evaluate hitters. Here we list some of the measures and they will be carefully compared and evaluated in later chapters. Since OBP is a measure of a player's ability to get on-base and SLG is a measure of a player's ability to advance the runners, a simple measure OPS (for **O**n-base percentage **P**lus **S**lugging percentage), adds the two measures:

$$\text{OPS} = \text{OBP} + \text{SLG}.$$

In the chapters to follow, we will investigate the goodness of the measures RC and OPS in predicting the number of runs scored by a team.

Baseball Data

Here we describe several common types of baseball data that we will analyze.

Career Statistics for a Player. Probably the most familiar data set among baseball fans is the batting or pitching statistics of a player over the seasons of his career. Many fans have collected baseball cards, either as children or adults, and these cards typically contain these career statistics. Figure 1.4 shows the front and back sides of a 2001 Topps card for Jim Thome, who played for the Cleveland Indians. For each of the baseball seasons from 1988 through 2000, the card gives the number of games (G) played by Thome, the at-bats (AB), runs scored (R), hits (H), doubles (2B), triples (3B), home runs (HR), runs batted in (RBI), stolen bases (SB), slugging percentage (SLG), walks (BB), strikeouts (SO), and the batting average (AVG).

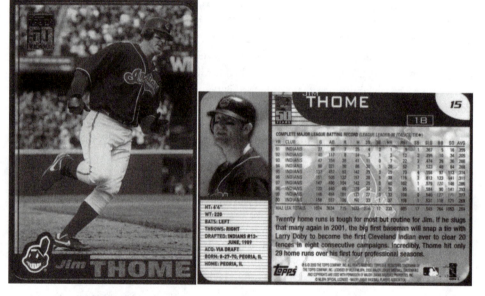

FIGURE 1.4
Front and back sides of the 2001 Topps baseball card for Jim Thome.

Player Statistics for a Given Season. Another informative data set is the collection of all batting and pitcher statistics for all players in a particular season. We will see that it is difficult to judge a single statistic, say Roger Maris's 61 home-run season in 1961, by itself. (Roger Maris is famous since he set the single season record for home runs that particular year.) To understand the significance of this statistic, we need to look at it in the context of all player statistics for that season. Although the basic rules of baseball have not changed over the years, the athletic abilities of the players, the competitive balance, and the equipment and environment have changed, and these changes have had a substantive impact on the values of baseball statistics. To return to our Johnny Pesky and Dick Stuart comparison, it really is not fair to compare the runs created by Pesky and Stuart at face value, since there were many more runs scored in the 1961 season than the 1942 season.

Team Statistics for a Given Season. Another interesting data set to work with is the batting and pitching statistics for all teams in a particular season. One general problem of interest is to find a suitable measure of the hitting ability of a player. Since the goal of batting is to score runs, one is interested in finding a hitting statistic that is useful in predicting the number of runs scored. But teams, not individual players, score runs, so one needs to look at team data in the development of good hitting measures.

Game Logs. Baseball fans are fascinated with the day-to-day performance of their favorite players and teams. Teams and individual players go through good and bad periods and these performances in short time periods are often described in the media. So it is interesting to look at the performance of players and teams for each game of the baseball season.

Situational Statistics. Baseball fans are also fascinated with the performances of teams and players in given situations. How does a batter perform at home and away games? How does he perform against different pitchers? How does he do at night and day games? How well does he hit when he swings at the first pitch, or when there are two strikes in the count? How well does a team perform when their ace pitcher is starting? These situational or breakdown statistics are now reported on all of the popular baseball news websites. One goal of this book is to try to make sense of the importance of these statistics.

Statistics Through the Years. Baseball has a fascinating history. It is fun to read about the great players of the past, like Ty Cobb, Walter Johnson, Shoeless Joe Jackson, Joe DiMaggio, and Babe Ruth. Also it is interesting to look at the great teams of all time, including the 1927 New York Yankees, the 1929 Philadelphia Athletics, the 1975 Cincinnati Reds, and the 1998 New York Yankees. One can explore this baseball history by means of baseball statistics. For example, we will look at home runs hit by teams in the years 1927, 1961, 1998, and 2001, and the differences that we see in this comparison tell us a lot about the relative difficulty of hitting a home run over time.

Miscellaneous Statistics. Although we will focus much of our discussion on basic hitting and pitching statistics, there are many other associated baseball statistics that are fun to explore. These include:

- The salaries of the players
- The attendance counts at the games
- Statistics related to managerial strategy
- The duration of the game
- The number of pitches thrown
- The time needed to complete the game

Collecting Baseball Data

The collection of baseball data is much easier today than in the past due to the Internet. Many Internet web sites contain historical and current baseball data. Here we highlight several Internet sources that are convenient for downloading data that can be easily entered into a statistical computing package.

`www.sportsline.com`. The CBS Sportsline site is typical of the many baseball news Internet sites that are available. If you click on MLB, then Stats, then a number of baseball data sets are available for viewing. This site gives career statistics for a current player, team statistics, and many types of situational statistics for teams and players. One difficulty of this site is currently that most of the data are available in html tables that are awkward to download for use in statistics packages. This site is good for finding leaders in particular statistical categories and for comparing teams and players.

`www.baseball-reference.com`. This Internet site is a good site for historical data on teams and players. One can easily obtain career statistics for any historical player and team data is available for any past season. The data sets are stored as text files and it is relatively easy to import the data into a standard spreadsheet program such as *Microsoft Excel*.

`www.baseball1.com`. The Baseball Archive site is an especially good site for downloading large collections of baseball data. In fact, one can download (either in text or Microsoft Access format) a single file that contains player and team statistics for all years in baseball history. Many of the data sets used in this book are taken from this database.

`www.retrosheet.org`. The Retrosheet organization is dedicated to collecting play-by-play baseball data. For each plate appearance in a given game, a data file will record the name of the hitter, the name of the pitcher, the game situation (score, runners on base, number of outs), the play, and other information. One can currently download this type of data for entire baseball seasons. Using these data, one can perform many interesting analyses. This type of data will be used in the Markov Chain modeling described in Chapter 9.

2

Exploring a Single Batch of Baseball Data

What's On-Deck?

In this chapter, we illustrate a number of graphs and summary statistics useful in exploring a single batch of data. In Case Study 2-1, we begin with batting statistics for the 30 Major League Teams for the 2001 season. We focus on the team home run numbers and use stemplots and five-number summaries to compare the home run production of the National League and American League teams. In the next two case studies, we look at the career statistics for two future Hall of Famers, Cal Ripken and Roger Clemens. When looking at an individual's statistic, it is helpful to construct a graph of the statistic against time. The patterns in this time series plot are helpful for understanding how the player has matured as a baseball player. In Case Study 2-4, we study baseball attendance for the 30 teams in 2001. Although baseball teams would all like to make a profit, we will see a wide disparity of the teams' abilities to bring fans to the ballpark. We conclude in Case Study 2-5 by looking at statistics for managers. One basic play in baseball is the sacrifice bunt, and we will see that this is a popular strategy for some managers and a very unpopular move for other managers.

Case Study 2-1: Looking at Teams' Offensive Statistics

Topics Covered: Stemplot, data distribution, five-number summary   After the conclusion of a baseball season in November, teams begin to evaluate how well they performed during the season. How effective was a particular team, say the Phillies, in getting batters on base? Did the Phillies score a lot of runs this year? What teams were good and bad in hitting home runs? How many home runs were hit by a representative major league team this year?

Table 2.1 displays a number of offensive statistics for all 30 major league teams for the 2001 season. Many of these statistics are *counts*, such as the number of hits, the number of doubles, the count of home runs, and so on. Other offensive statistics are *derived measures*

TABLE 2.1

Batting statistics for all Major League Baseball teams in 2001.

American League

Team	G	Avg	OBP	Slg	AB	R	H	2B	3B	HR	RBI	BB	SO	HBP
Anaheim (ANA)	162	.261	.327	.405	5551	691	1447	275	26	158	662	494	1001	77
Baltimore (BAL)	161	.248	.319	.380	5472	687	1359	262	24	136	663	514	989	77
Boston (BOS)	161	.266	.334	.439	5605	772	1493	316	29	198	739	520	1131	70
Chicago (CHW)	162	.268	.334	.451	5464	798	1463	300	29	214	770	520	998	52
Cleveland (CLE)	162	.278	.350	.458	5600	897	1559	294	37	212	868	577	1076	69
Detroit (DET)	162	.260	.320	.409	5537	724	1439	291	60	139	691	466	972	51
Kansas City (KC)	162	.266	.318	.409	5643	729	1503	277	37	152	691	406	898	44
Minnesota (MIN)	162	.272	.337	.433	5560	771	1514	328	38	164	717	495	1083	64
New York (NYY)	161	.267	.334	.435	5577	804	1488	289	20	203	774	519	1035	64
Oakland (OAK)	162	.264	.345	.439	5573	884	1469	334	22	199	835	640	1021	88
Seattle (SEA)	162	.288	.360	.445	5680	927	1637	310	38	169	881	614	989	62
Tampa Bay (TB)	162	.258	.320	.388	5524	672	1426	311	21	121	645	456	1116	54
Texas (TEX)	162	.275	.344	.471	5685	890	1566	326	23	246	844	548	1093	75
Toronto (TOR)	162	.263	.325	.430	5663	767	1489	287	36	195	728	470	1094	74

National League

Team	G	Avg	OBP	Slg	AB	R	H	2B	3B	HR	RBI	BB	SO	HBP
Arizona (AZ)	162	.267	.341	.442	5595	818	1494	284	35	208	776	587	1052	57
Atlanta (ATL)	162	.260	.324	.412	5498	729	1432	263	24	174	696	493	1039	45
Chicago (CHI)	162	.261	.336	.430	5406	777	1409	268	32	194	748	577	1077	66
Cincinnati (CIN)	162	.262	.324	.419	5583	735	1464	304	22	176	690	468	1172	65
Colorado (COL)	162	.292	.354	.483	5690	923	1663	324	61	213	874	511	1027	61
Florida (FLA)	162	.264	.326	.423	5542	742	1461	325	30	166	713	470	1145	67
Houston (HOU)	162	.271	.347	.451	5528	847	1500	313	29	208	805	581	1119	89
Los Angeles (LA)	162	.255	.323	.425	5493	758	1399	264	27	206	714	519	1062	56
Milwaukee (MIL)	162	.251	.319	.426	5488	740	1378	273	30	209	712	488	1399	72
Montreal (MON)	162	.253	.319	.396	5379	670	1361	320	28	131	622	478	1071	60
New York (NY)	162	.249	.323	.387	5459	642	1361	273	18	147	608	545	1062	65
Philadelphia(PHI)	162	.260	.329	.414	5497	746	1431	295	29	164	708	551	1125	43
Pittsburgh (PIT)	162	.247	.313	.393	5398	657	1333	256	25	161	618	467	1106	67
San Diego (SD)	162	.252	.336	.399	5482	789	1379	273	26	161	753	678	1273	41
San Francisco(SF)	162	.266	.342	.460	5612	799	1493	304	40	235	775	625	1090	50
St. Louis (STL)	162	.270	.339	.441	5450	814	1469	274	32	199	768	529	1089	65

of offensive performance, such as batting average (AVG), slugging percentage (SLG), and on-base percentage (OBP), which are computed from the count statistics.

Home Run Totals. Since there were a lot of home runs hit in 2001 (including Barry Bonds' record-setting 73), let's look at the number of home runs hit by the 30 teams.

The first step in exploring a single batch of data, such as these 30 home run totals, is to draw a suitable graph. An effective graph that is easy to draw by hand is the stemplot. To construct a stemplot, we divide each home run total into two parts, called the stem and the leaf. For example, Anaheim's home run total, 158, can be divided between the tens and units places (see below)

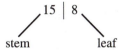

to get a stem of 15 and a leaf of 8. We write down all of the possible stems, and record each home run total by writing down its leaf on the line corresponding to the stem. If we do this for all 30 home run totals, we get the stemplot shown in Figure 2.1.

```
12 | 1
13 | 169
14 | 7
15 | 28
16 | 114469
17 | 46
18 |
19 | 45899
20 | 36889
21 | 234
22 |
23 | 5
24 | 6
```

FIGURE 2.1
Stemplot of team home run numbers from 2001 season.

The second line tells us that the totals 131, 136 and 139 were the number of home runs hit by three of the 30 teams. To study this distribution of home run totals, it may be helpful to flip this stemplot by a 90-degree turn, so that the small totals are on the left.

```
    1  169  7  28  114469  46     45899  36889  234        5  6
    12 13   14 15 16      17  18  19     20     21  22  23 24
```

What observations can we make from this data distribution?

1. First, we look for the general *shape* of these home run totals. Here we see two clusters of totals—one between 152 and 176 and a second between 194 and 214. In-

terestingly, there is a gap between these two clusters, so it is easy to categorize the teams as hitting "low" and "high" numbers of home runs.

2. After we think of the general shape, we look for an "average" home run total. Since exactly half (15) of the totals fall below 180 and half fall above 180, we can regard 180 home runs as a measure of the center of the distribution. (We call 180 the *median* of the observations.)

3. Next, we look at the *spread* or variation in these home run totals. Here the spread of the totals is pretty large, the largest value 246 (number of home runs hit by Texas) is more than twice as large as the smallest value 121 (hit by Tampa Bay). But most of the home run totals fall in the two clusters between 152 and 214.

4. Last, we look for any *unusual* characteristics of the totals. We've already discussed the gap in the middle of the dataset. Also there are two large numbers that are separated from the rest—these two teams (San Francisco and Texas) hit a lot of home runs in 2001.

If you are a baseball fan, you are probably interested in the relative standing of your team in this distribution of home run totals. To better see the teams' relative standing, we can redraw this stemplot in Figure 2.2 using team labels (given in the table above) instead of numerical leafs. (The National League teams are indicated by bold type.)

```
12 | TB
13 | MON BAL DET
14 | NY
15 | KC ANA
16 | PIT SD PHI MIN FLA SEA
17 | ATL CIN
18 |
19 | CHI TOR BOS OAK STL
20 | NYY LA MIL HOU AZ
21 | CLE COL CHW
22 |
23 | SF
24 | TEX
```

FIGURE 2.2
Stemplot of team home run numbers from 2001 season with teams identified.

My team, the Phillies, appear in the cluster of lower numbers. It is interesting to note that home run production is not always associated with winning and losing. Texas, the team with the highest home run total, finished in last place in their division, whereas Seattle, a team with one of the best win/loss records of all time, had a team home run total in the bottom half of the distribution.

This stemplot display motivates a follow-up question: What league hit more home runs in 2001? To compare the team totals for the two leagues, in Figure 2.3 we will put the leaves of the home run totals for the NL teams to the left of the stems and the leaves for the home run totals for the AL on the right.

	National League				American League
			12	TB	
	MON		13	BAL DET	
	NY		14		
			15	KC ANA	
FLA PHI SD PIT			16	MIN SEA	
	CIN ATL		17		
			18		
	STL CHI		19	TOR BOS OAK	
AZ HOU MIL LA			20	NYY	
	COL		21	CLE CHW	
			22		
	SF		23		
			24	TEX	

FIGURE 2.3
Back-to-back stemplots of team home run numbers from the National and American Leagues.

Comparing the left and right stemplots, it appears that the two leagues had approximately the same number of home runs in 2001. Note that exactly eight of the 16 National League teams and seven of the 14 American League teams hit more than 180 home runs. This result might surprise you since the American League plays with the *designated hitter*, who is a substitute hitter for the pitcher (in contrast to the National League where the pitcher comes to bat).

On-Base Percentages. Actually, the number of home runs hit is not really a good reflection of a team's offensive production. A better indicator of run production is the team's on-base percentage (OBP), which measures the fraction of plate appearances in which the team gets on-base. In constructing a stemplot of the 30 team OBPs, we break the OBP between the 2nd and 3rd digits. Also we write two lines for each possible stem, where the leaves 0–4 are written on the first line and the leaves 5–9 on the second line. The second line tells us that one team had an on-base percentage of .318 and three teams had an on-base percentage of .319. In Figure 2.4 we show the basic stemplot on the left, and show this stemplot rotated 90 degrees in the center. To show the basic pattern, we draw a smooth curve on top of the stemplot on the right.

What do we see in these displays?

1. The shape of these team OBPs is roughly right-skewed with a long tail to the right. There are a large number of OBPs in the .318–.329 range and the shape of the stemplot decreases slowly as the OBP gets larger.

2. An average OBP is about .325 and the team OBPs range from .313 (Pittsburgh) to .360 (Seattle).

3. One unusual feature that stands out is the gap between Seattle's OBP of .360 and the next two team OBPs. Evidently, one reason why Seattle had such a great season in 2001 was their batters' high rate of getting on-base.

Basic Display	Rotated 90 Degrees	Smooth Curve Drawn on Top
31 │ 3		
31 │ 8999		
32 │ 003344		
32 │ 5679		
33 │ 444		
33 │ 6679		
34 │ 124		
34 │ 57		
35 │ 04		
35 │		
36 │ 0		

FIGURE 2.4

Stemplot of the OBPs for the 2001 Major League teams.

One can summarize these team OBPs by the computation of a median and quartiles. The *median* is the value that divides the data into a bottom half and a top half. Here we have 30 values—so the median is the average of the 15th and 16th largest OBPs:

$$\text{Median} = (.329 + .334)/2 = .332.$$

The quartiles divide the data into quarters. The *lower quartile* is the median of the lower half of the data, and the *upper quartile* is the median of the upper half. In our example, we can divide the team OBPs into an upper half of 15 values, and a lower half of 15 values. The median of the lower half is the 8th lowest value (.323), and the median of the upper half is the 8th largest value (.341). So

$$\text{lower quartile} = .323, \quad \text{upper quartile} = .341.$$

A five-number summary of these data is (lowest value, lower quartile, median, upper quartile, highest value), which is (.313, .323, .332, .341, .360). These five numbers divide the data roughly in quarters. So approximately $\frac{1}{4}$ or 25% of the team OBPs fall between .313 and .323, approximately 25% of the OBPs fall between .323 and .332, and so on.

Again, it's interesting to note the relative standing of the OBP of your favorite team. My team (the Phils) had a team OBP of .329, that puts it in the lower half of the dataset. My prediction was the Phillies would try to improve their offensive production during the 2001–2 off-season by signing free agents or getting players by trade who appear to be effective in getting on-base. (The reader can verify from the records the validity of that prediction.)

Case Study 2-2: A Tribute to Cal Ripken

Topics Covered: Dotplot, stemplot, time series plot, fitted line   The 2001 baseball season was memorable for the retirement of one of baseball's most popular players, Cal Ripken. Cal played infield for the Baltimore Orioles for 21 years. He is best known for his durability. He set the record for the most consecutive games played—he played in 2632 consecutive Orioles games between 1982 and 1998.

Here are Cal's batting statistics for all of his major league seasons. We'll focus on a couple of interesting statistics—the number of home runs (HR) and the OPS statistic that is a good estimate of a player's overall hitting ability. In the following, we will not include the statistics for his rookie 1981 season since he only had 39 at-bats, and we'll analyze his data for the remaining 20 seasons.

TABLE 2.2
Batting statistics for Cal Ripken's career.

YR	G	AB	R	H	2B	3B	HR	RBI	TB	AVG	OBP	SLG	OPS
1981	23	39	1	5	0	0	0	0	5	.128	.150	.128	.278
1982	160	598	90	158	32	5	28	93	284	.264	.317	.475	.792
1983	162	663	121	211	47	2	27	102	343	.318	.371	.517	.888
1984	162	641	103	195	37	7	27	86	327	.304	.374	.510	.884
1985	161	642	116	181	32	5	26	110	301	.282	.347	.469	.816
1986	162	627	98	177	35	1	25	81	289	.282	.355	.461	.816
1987	162	624	97	157	28	3	27	98	272	.252	.333	.436	.769
1988	161	575	87	152	25	1	23	81	248	.264	.372	.431	.803
1989	162	646	80	166	30	0	21	93	259	.257	.317	.401	.718
1990	161	600	78	150	28	4	21	84	249	.250	.341	.415	.756
1991	162	650	99	210	46	5	34	114	368	.323	.374	.566	.940
1992	162	637	73	160	29	1	14	72	233	.251	.323	.366	.689
1993	162	641	87	165	26	3	24	90	269	.257	.329	.420	.749
1994	112	444	71	140	19	3	13	75	204	.315	.364	.459	.823
1995	144	550	71	144	33	2	17	88	232	.262	.324	.422	.746
1996	163	640	94	178	40	1	26	102	298	.278	.341	.466	.807
1997	162	615	79	166	30	0	17	84	247	.270	.331	.402	.733
1998	161	601	65	163	27	1	14	61	234	.271	.331	.389	.720
1999	86	332	51	113	27	0	18	57	194	.340	.368	.584	.952
2000	83	309	43	79	16	0	15	56	140	.256	.310	.453	.763
2001	128	477	43	114	16	0	14	68	172	.239	.276	.361	.637

Cal's Home Runs. Scanning over Cal's hitting statistics, we see that he displayed some power and hit a good number of home runs in his career. We graph his season-by-season home runs using a dotplot in Figure 2.5.

There appears to be a lot of variation in these home run numbers—we see from the graph that the numbers range from 13 to 34. There is one cluster of values in the 13–15 range, and another cluster in the 26–28 range. The median number of home runs hit is 22 but Cal did not hit 22 home runs for any season.

Maybe some of the variation of Cal's home run numbers can be explained by the age at which he hit them. A ballplayer generally improves in hitting ability in the early part of

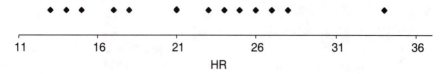

FIGURE 2.5
Dotplot of season home run numbers for Cal Ripken.

his career and declines in ability toward the end of his career. In Figure 2.6, we graph the home run count against the year.

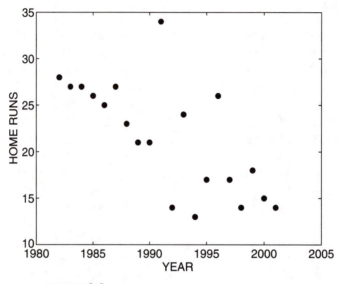

FIGURE 2.6
Time series plot of home run numbers for Cal Ripken.

We see that Cal's home run numbers were in the 25–30 range during the years 1982–1987, but it appears that his season home run counts generally decreased over the later years of his career. We can summarize this decrease by drawing a line through the points in Figure 2.7.

The equation of this line is

$$\text{home run} = 1446 - 0.715 \text{ year.}$$

So Cal's home run count tended to decrease about 0.7 for each year of his career. There were some exceptions to this general pattern, such as his 34 home runs hit in 1991 and his 26 home runs in 1996.

Cal's Season OPS Values. Although Cal's home runs decreased during his career, it's not clear that his hitting ability changed in a similar way. After all, good hitting is more than

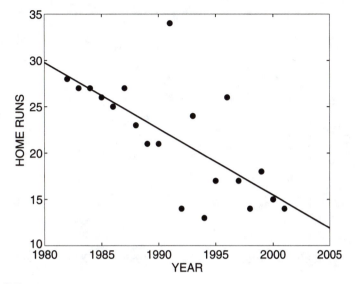

FIGURE 2.7
Time series plot of home run numbers for Cal Ripken with a "good fitting" line drawn on top.

just hitting home runs. As explained in Chapter 1, a good estimate of hitting effectiveness is the OPS statistic. In Figure 2.8 we display Cal's season OPS values for his 20 seasons using a stemplot.

```
6 | 3
6 | 8
7 | 12344
7 | 5669
8 | 00112
8 | 88
9 | 4
9 | 5
```

FIGURE 2.8
Stemplot of Ripken's season OPS values.

From the stemplot, we see that Cal's OPS values are roughly bell-shaped centered about .750. Most of his season values fall between .710 and .880 with a few extreme values. To see how Cal's OPS values changed over the seasons, we construct the scatterplot in Figure 2.9.

We don't see the strong decreasing pattern across time that we saw in the home run counts. Cal's OPS values generally fell between .7 and .8. He had two strong hitting years in the early part of his career, another good year in 1991, and his best year in 1999. Generally this figure reinforces the impression that Cal was a consistently good ball player during his whole career.

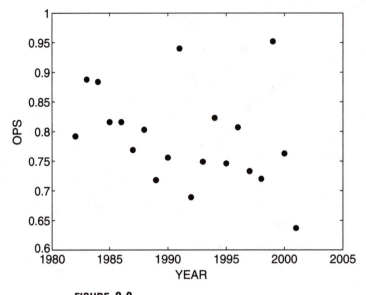

FIGURE 2.9
Time series plot of Cal Ripken's OPS numbers.

Case Study 2-3: A Tribute to Roger Clemens

Topics Covered: Stemplot, time series plot, summary statistics, comparison of distributions   Roger Clemens has been one of the greatest strikeout pitchers in modern baseball. He has a lifetime (through the 2001 season) win/loss record of 280–145 and he has won the Cy Young award as the best pitcher in the league for the 1986, 1987, 1991, 1997, 1998 and 2001 seasons. The following table gives Roger's pitching statistics for his first 18 seasons in Major League Baseball. One should note an unusual feature of the innings pitched (IP) data. The number 133.1 denotes 133 and $\frac{1}{3}$ innings pitched, and 281.2 denotes 281 and $\frac{2}{3}$ innings pitched.

Roger's Strikeouts. The strikeout numbers, 126, 74, and so on, are hard to interpret since the number of innings pitched changes across seasons. A reasonable measure of strikeout ability that adjusts for the number of innings is the *strikeout rate*, defined by

$$\text{Strikeout rate} = \frac{\text{SO}}{\text{IP}} \times 9.$$

If we divide the count of strikeouts by the innings pitched, we get the number of strikeouts per inning. By multiplying the ratio SO/IP by 9, we get the number of strikeouts for a standard 9-inning game. A strikeout rate of 9 is a useful reference value, since it means that the pitcher struck out a batter per inning. (This particular rate value is quite rare.) Since there are 27 outs for a team during a game, a strikeout rate of 9 means that one third of the outs were strikeouts. If we compute the strikeout rate for all of Roger's seasons, we get the table shown in Figure 2.10. A stemplot of the rates is shown to the right of the table.

TABLE 2.3
Pitching statistics for Roger Clemens.

YR	TEAM	LG	W	L	PCT	G	SV	IP	H	R	ER	SO	TBB	IBB	ERA
1984	Bos	AL	9	4	.692	21	0	133.1	146	67	64	126	29	3	4.32
1985	Bos	AL	7	5	.583	15	0	98.1	83	38	36	74	37	0	3.29
1986	Bos	AL	24	4	.857	33	0	254.0	179	77	70	238	67	0	2.48
1987	Bos	AL	20	9	.690	36	0	281.2	248	100	93	256	83	4	2.97
1988	Bos	AL	18	12	.600	35	0	264.0	217	93	86	291	62	4	2.93
1989	Bos	AL	17	11	.607	35	0	253.1	215	101	88	230	93	5	3.13
1990	Bos	AL	21	6	.778	31	0	228.1	193	59	49	209	54	3	1.93
1991	Bos	AL	18	10	.643	35	0	271.1	219	93	79	241	65	12	2.62
1992	Bos	AL	18	11	.621	32	0	246.2	203	80	66	208	62	5	2.41
1993	Bos	AL	11	14	.440	29	0	191.2	175	99	95	160	67	4	4.46
1994	Bos	AL	9	7	.563	24	0	170.2	124	62	54	168	71	1	2.85
1995	Bos	AL	10	5	.667	23	0	140.0	141	70	65	132	60	0	4.18
1996	Bos	AL	10	13	.435	34	0	242.2	216	106	98	257	106	2	3.63
1997	Tor	AL	21	7	.750	34	0	264.0	204	65	60	292	68	1	2.05
1998	Tor	AL	20	6	.769	33	0	234.2	169	78	69	271	88	0	2.65
1999	NYY	AL	14	10	.583	30	0	187.2	185	101	96	163	90	0	4.60
2000	NYY	AL	13	8	.619	32	0	204.1	184	96	84	188	84	0	3.70
2001	NYY	AL	20	3	.870	33	0	220.1	205	94	86	213	72	1	3.51

YEAR	STRIKEOUT RATE
1984	8.50
1985	6.77
1986	8.43
1987	8.18
1988	9.92
1989	8.17
1990	8.24
1991	7.99
1992	7.59
1993	7.51
1994	8.86
1995	8.49
1996	9.53
1997	9.95
1998	10.39
1999	7.82
2000	8.28
2001	8.70

STEMPLOT

6	7
7	
7	5589
8	112244
8	578
9	
9	599
10	3

FIGURE 2.10
Table and stemplot of season strikeout rates for Roger Clemens.

Looking at the stemplot, we see that an average strikeout rate for Roger is about 8.2. That means that he strikes out almost one batter per inning. For most of his seasons, Roger's strikeout rate was in the 7.5–8.8 range, although there were four seasons where his rate exceeded 9. To see when these strong strikeout seasons occurred, we plot the rates against the season year in Figure 2.11. We place a horizontal line on our graph corresponding to the reference strikeout rate of 9.0.

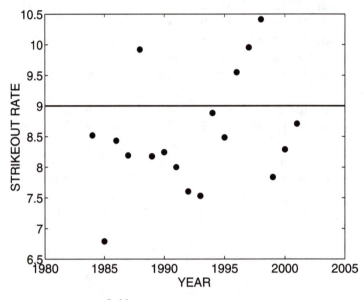

FIGURE 2.11
Time series plot of Clemen's season strikeout rates.

There is an interesting pattern in this plot of strikeout rates. From 1984 to 1993, Roger's strikeout ability seemed to generally go down, but after 1994, his strikeout rate went up, reaching a peak at 1998. When Roger moved to the Yankees, his strikeout rate dropped under 8.0 but increased the next two years. It is not clear if this pattern in the plot corresponds to any change in Roger's physical condition or his style of pitching, but it may deserve further investigation.

Roger's Win/Loss Record in 2001. Roger Clemens had an amazing win/loss record in 2001—he won 20 and lost only three games. For this reason (and others), Clemens won the Cy Young award for the best pitcher in the American League in 2001. But was Clemens really the best AL pitcher in 2001?

One problem with a pitcher's win/loss record is that winning and losing is really a team accomplishment. A team will win games if it scores more runs than its opponent. This requires good pitching that allows few runs and good hitting that produces runs. It is possible that Clemens won so many games in 2001 because his team, the Yankees, scored a lot of runs when he was pitching. To see if this is true, we record the number of runs scored by the Yankees for every one of the 34 games that Clemens started in the 2001 season. The run numbers are shown below. We see that the Yankees scored seven runs in the first game that Clemens started, 16 runs in the second game he started, and so on.

7	16	3	6	5	3	6	2	2	2
12	9	4	9	10	2	7	4	5	8
7	12	5	4	10	8	7	3	4	6
0	1	4							

We display the runs scored for Clemens in Figure 2.12 using a stemplot.

```
0 | 01
0 | 2222333
0 | 44444555
0 | 6667777
0 | 8899
1 | 00
1 | 22
1 | 6
```

FIGURE 2.12
Stemplot of runs scored by the Yankees during games in which Clemens started in 2001.

We see that the distribution of runs scored is right skewed. We see that most of the runs scored are in the 2–7 range—it was relatively unusual for the Yankees to score 0 or 1 runs, or to score more than 7 runs. The median runs scored for Clemens was 5 and the mean was 5.85 runs.

How does this distribution compare to the runs scored by the Yankees in the 2001 games where Clemens was not a starter? Figure 2.13 displays back-to-back stemplots of the runs scored for the two groups of games.

Comparing the two stemplots in Figure 2.13, we see that the Yankees tended to score fewer runs in games in which Clemens did not start than they did in games in which Clemens started. The median and mean runs scored for Yankee games when Clemens was not starting are 4 and 4.80, respectively. If we compare medians or means of the two groups, we see that the Yankees tended to score one more run than average when Clemens was pitching. So Clemens did receive good run support from his team.

Roger Clemens won the Cy Young award largely due to his great 20-3 win/loss record. But our work demonstrates that the Yankee hitters should receive some of the credit for Clemens's great win/loss record. Perhaps having Clemens as a starter gave his teammates a psychological boost.

Case Study 2-4: Analyzing Baseball Attendance

Topics Covered: Histogram, stemplot, data distribution   Baseball is a business. Each of the 30 major league teams are privately owned and each wants to make a profit. Much of the revenue produced by the teams comes from the money that is made from ticket and concession sales. So it is desirable for a team to have high attendances at the games played

CLEMENS WAS STARTER		CLEMENS WAS NOT STARTER
0	0	0000
0	1	000000000000
0000	2	000000000000000000000000
000	3	00000000000000
00000	4	0000000000000000000
000	5	000000000000
000	6	0000000000000
0000	7	000000000000
00	8	00000
00	9	0000000
00	10	0
	11	0
00	12	0
	13	0
	14	000
	15	0
0	16	0

FIGURE 2.13

Back-to-back stemplots of runs scored by the 2001 Yankees for games when Roger Clemens was the starter and games when Clemens was not the starter.

at its ballpark. We will see in this study that some teams are very successful and other teams are not successful in getting fans to come to their games.

Table 2.4 shows the mean attendance at the home ballpark for all thirty major league teams in 2001. We note that the teams who played in the 2001 World Series, Arizona (NL) and New York (AL), had mean attendances of 33,783 and 40,811 per game, respectively. Are these large or small values? We can answer this question by looking at the distribution of mean attendances for all teams.

We first construct a histogram of the attendances for all teams. We chose to group the data using the bins (7000, 12,000], (12,000, 17,000], ... , (42,000, 47,000] and count the number of values in each bin. The graph of the grouped data, called a histogram, is shown in Figure 2.14.

There are a number of interesting features about these data that we see from the histogram.

- There is a large range in the average attendances, from about 7000 to 47,000.

- The shape of this distribution is bimodal—there are two clumps of attendance values. Eight of the thirty teams had average attendances between 22,000 and 27,000, and 15 of the teams had attendances between 32,000 and 42,000.

- Four teams had small average attendances under 22,000.

TABLE 2.4

Mean home attendance for all Major League Baseball teams in 2001.

2001 American League		2001 National League	
Team	Home Avg.	Team	Home Avg.
Anaheim (ANA)	24702	Arizona (AZ)	33783
Baltimore (BAL)	38661	Atlanta (ATL)	34858
Boston (BOS)	32413	Chicago (CHI)	35183
Chicago (CHI)	22077	Cincinnati (CIN)	23794
Cleveland (CLE)	39694	Colorado (COL)	39019
Detroit (DET)	24087	Florida (FLA)	15765
Kansas City (KC)	18967	Houston (HOU)	35855
Minnesota (MIN)	22286	Los Angeles (LA)	37248
New York (NYY)	40811	Milwaukee (MIL)	34704
Oakland (OAK)	26336	Montreal (MON)	7935
Seattle (SEA)	43300	New York (NY)	32818
Tampa Bay (TB)	16029	Philadelphia (PHI)	22846
Texas (TEX)	34950	Pittsburgh (PIT)	30806
Toronto (TOR)	23647	San Diego (SD)	29726
		San Francisco(SF)	40888
		St. Louis (STL)	38389

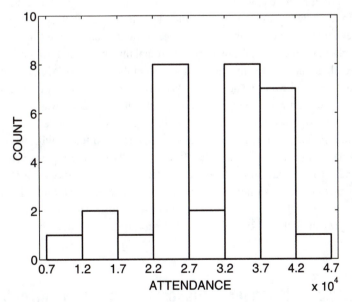

FIGURE 2.14

Histogram of average attendances for the Major League teams in 2001.

0	7	0	**MON**
0		0	
1		1	
1		1	
1	5	1	**FLA**
1	6	1	TB
1	8	1	KC
2		2	
2	22233	2	CHW MIN **PHI** TOR **CIN**
2	44	2	ANA DET
2	6	2	OAK
2	9	2	**SD**
3	0	3	PIT
3	223	3	BOS **NY AZ**
3	44455	3	TEX **ATL MIL CHI HOU**
3	7	3	**LA**
3	8899	3	BAL **STL** CLE COL
4	00	4	NYY **SF**
4	3	4	SEA

FIGURE 2.15

Stemplot of average attendances for the Major League teams in 2001; in the right display, the stem-plot leaves are replaced by team labels.

To look further at these teams' attendances, we construct a stemplot. In Figure 2.15, the basic display is shown below on the left, and the right display replaces the leaves with the team abbreviations. As before, the National League abbreviations are shown in bold face.

The graphs in Figure 2.15 give us some additional insight about the attendance averages. The four weak attendance teams are, from low to high, Montreal (MON), Florida (FLA), Tampa Bay (TB), and Kansas City (KC). Baseball attendance seems to be low in Florida and Canada. The team with the highest average attendance in 2001 was Seattle, which also had the best win/loss record that year. But winning games and attendance don't always go hand in hand. Philadelphia (PHI) and Oakland (OAK), two teams with winning records, had relatively small average attendances. In contrast, two losing teams, Baltimore (BAL) and Colorado (COL), had high attendances—these high attendances might be explained by the fact that both teams play in relatively new ballparks. It would be interesting for a future study to look into the impact of a number of variables, such as the population of the surrounding region and the newness of the ballpark, on the average home attendance of a baseball team.

Case Study 2-5: Manager Statistics: the Use of Sacrifice Bunts

Topics Covered: Dotplot, comparison of distributions   In this chapter, we've looked at statistics for teams and for individual players. Here we'll look at statistics for baseball managers. This might surprise you—how can we measure characteristics of a manager?

Managers do have different styles or strategies that they use in managing a game, and we can contrast managers by measuring how often particular strategies are used.

Here we focus on the use of the sacrifice bunt, one of the more popular strategic plays in baseball. Suppose there is a runner on first base with 0 or 1 out. Then the manager may elect to have the batter bunt, which is a short hit in the infield. If the bunt is successful, the runner on first advances to second. If the batter is out, this play is called a sacrifice bunt because the batter is sacrificing his at-bat to advance the runner to second. Is the sacrifice

TABLE 2.5
Sacrifice bunt statistics for all Major League Baseball managers in 2000.

League	Manager	Attempts	Success Rate	Favorite Inning
American	Fregosi	45	75.6	1
American	Garner	58	79.3	9
American	Hargrove	36	80.6	8
American	Howe	40	77.5	6
American	Kelly	37	75.7	3
American	C. Manual	59	84.7	1
American	J. Manuel	75	86.7	7
American	Muser	72	87.5	7
American	Oates	66	83.3	8
American	Pinella	73	87.7	5
American	Rothsfield	73	79.5	7
American	Scioscia	63	84.1	7
American	Torre	22	81.8	3
American	Williams	49	89.8	7
National	Alou	103	83.5	3
National	Baker	86	90.7	3
National	Baylor	115	84.3	3
National	Bell	100	83	4
National	Bochy	52	76.9	2
National	Boles	61	78.7	3
National	Cox	109	80.7	2
National	Dierker	77	79.2	4
National	Francona	89	84.3	3
National	Johnson	80	81.3	2
National	La Russa	107	79.4	2
National	Lamont	78	76.9	3
National	Lopes	78	78.2	3
National	McKeon	82	76.8	3
National	Showalter	89	75.3	3
National	Valentine	84	85.7	2

bunt a good strategy? It can be a good strategy when there is a relatively weak batter at the plate or the objective is to score a single run.

Major League Handbook 2001 by Stats Inc. gives the following sacrifice bunt statistics for all 30 Major League managers in 2000. Table 2.5 gives, for each team, the number of sacrifice bunts attempted all year, the percentage of time the bunt was successful, and the most common inning where this type of bunt occurred.

The dotplot in Figure 2.16 displays the number of sacrifice bunt attempts for all teams. The median number of "sac-bunts" was 74. But we note a wide spread—Joe Torre of the Yankees only attempted 22 sacrifice bunts all season and other managers sacrificed over 100 times (in a 162 game season).

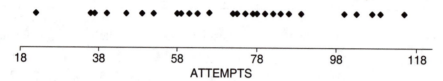

FIGURE 2.16
Dotplot of number of sacrifice bunt attempts by the Major League managers in the 2000 season.

How can we explain the wide spread in the number of sacrifice bunts among teams? Perhaps some managers don't think that the sacrifice bunt is a good strategy. Or possibly a manager of a team with strong pitching uses a sacrifice more often than a manager of a team with weak pitching.

Is it possible that the number of sacrifice bunts differs between National League and American League teams? We can check this by constructing two parallel dotplots in Figure 2.17—one for the NL teams and a second for the AL teams.

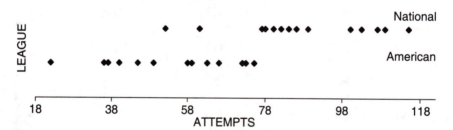

FIGURE 2.17
Parallel dotplots of number of sacrifice bunt attempts by National and American League managers.

We see in the Figure 2.17 display that there is quite a difference in the number of sacrifice bunts attempted in the two leagues. The median number of attempts for NL teams is 85 (about 1 in every 2 games) contrasted with 58.5 (about 1 in every 3 games) for the AL teams. There actually is one simple explanation for this discrepancy in sacrifice bunts—the designated hitter rule. Because National League pitchers (typically weak hitters) come to bat, it is common for them to attempt a sacrifice bunt when there is a runner on first base. They do this to avoid a double play and to advance the runner to second. In contrast, in the AL, the pitcher doesn't bat (the designated hitter does instead) and there is much less reason to use this strategy.

The league policy does explain many of the differences in attempts that we see. However, even within one league, we still see large differences in sacrifice bunt attempts. It is interesting to note that Joe Torre's 22 sac-bunt attempts is a small value even compared to other AL managers, and the NL manager values ranged between 52 and 115. Why do these differences exist? Well, the number of sacrifice bunts attempted depends partly on the batting abilities of the players and the beliefs of the manager regarding the value of the sacrifice bunt. It would be interesting to see if there is any relationship between the number of sacrifice bunts and other batting statistics. Also, one could see if particular managers have tendencies over many years to attempt a large or small number of sacrifice bunts.

Exercises

Leadoff Exercise. Table 2.6 displays the slugging percentages (SLGs) for Rickey Henderson for his first 23 years in the major leagues.

TABLE 2.6
Slugging percentages for Rickey Henderson for the first 23 years of his career.

AGE	SLG	AGE	SLG
20	.336	32	.423
21	.399	33	.457
22	.437	34	.474
23	.382	35	.365
24	.421	36	.447
25	.458	37	.344
26	.516	38	.342
27	.469	39	.347
28	.497	40	.466
29	.399	41	.305
30	.399	42	.351
31	.577		

(a) Construct a stemplot of the slugging percentages.

(b) Compute a five-number summary of the SLGs. What was a representative slugging percentage for Rickey?

(c) Construct a time series plot of the SLG against his age. Comment on any pattern that you see in this plot.

(d) Based on your work in (c), do you have any explanations for the low slugging percentages that Rickey had in his career?

2.1. Table 2.7 gives the career batting statistics for the great Yankee player Joe DiMaggio.

TABLE 2.7
Career statistics for Joe DiMaggio.

YR	TM	L	G	AB	R	H	HR	RBI	AVG	OBP	SLG
1936	NYY	A	138	637	132	206	29	125	.323	.352	.576
1937	NYY	A	151	621	151	215	46	167	.346	.412	.673
1938	NYY	A	145	599	129	194	32	140	.324	.386	.581
1939	NYY	A	120	462	108	176	30	126	.381	.448	.671
1940	NYY	A	132	508	93	179	31	133	.352	.425	.626
1941	NYY	A	139	541	122	193	30	125	.357	.440	.643
1942	NYY	A	154	610	123	186	21	114	.305	.376	.498
1946	NYY	A	132	503	81	146	25	95	.290	.367	.511
1947	NYY	A	141	534	97	168	20	97	.315	.391	.522
1948	NYY	A	153	594	110	190	39	155	.320	.396	.598
1949	NYY	A	76	272	58	94	14	67	.346	.459	.596
1950	NYY	A	139	525	114	158	32	122	.301	.394	.585
1951	NYY	A	116	415	72	109	12	71	.263	.365	.422
(13 seasons)			1736	6821	1390	2214	361	1537	.325	.398	.579

(a) Construct a dotplot of DiMaggio's yearly batting averages on the number line below.

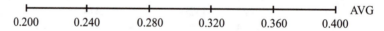

(b) Describe the basic features of this distribution (shape, average value, spread, and unusual values).

(c) What proportion of the years was DiMaggio at least a .300 hitter?

(d) Construct a dotplot of DiMaggio's yearly hits on the number line below.

(e) What was a median number of yearly hits for DiMaggio? Were there any unusually small or large hit values? What is a possible explanation for these unusual values?

2.2. (Continuation of Exercise 2.1)

    (a) Construct a time series plot of DiMaggio's AVG values on the graph below.

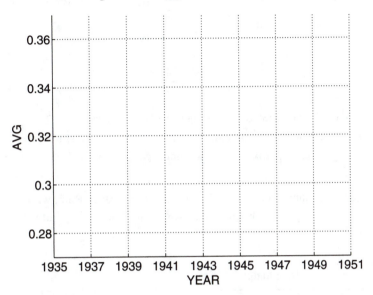

    (b) From the graph, DiMaggio appeared to peak in AVG two times during his career. Which two years?

    (c) From the graph, did you detect any gaps in DiMaggio's career? What is a possible explanation for these gaps?

    (d) Generally, ballplayers mature slowly and reach their maximum performance during the middle of their careers. Do you see any evidence for this maturation in DiMaggio's AVG plot? Explain.

    (e) Also, ballplayers generally decline in ability during the last part of their career. Do you see any decline in DiMaggio's AVG plot? Explain.

2.3. (Joe DiMaggio data from Exercise 2.1)

    (a) Construct a stemplot of DiMaggio's on-base percentages (OBP) for his 13 seasons.

    (b) Compute DiMaggio's mean OBP and his median OBP.

    (c) Compare the mean and median that you computed in (b). Which is a better measure of a representative season OBP for DiMaggio?

2.4. (Joe DiMaggio data from Exercise 2.1)   A stemplot of DiMaggio's season HR numbers is shown below:

```
1 | 24
1 |
2 | 01
2 | 59
3 | 00122
3 | 9
4 |
4 | 6
```

(a) Find a five-number summary of the HR numbers.

(b) Based on your calculations in part (a), half of all the HR numbers are smaller than ____ and approximately the middle 50 percent of the season HRs are between ____ and ____.

2.5. Table 2.8 shows the career hitting statistics for Babe Ruth, who was considered by ESPN to be the second greatest athlete of the 20th century (after Michael Jordan).

TABLE 2.8
Career statistics for Babe Ruth.

YR	TM	L	G	AB	R	H	HR	RBI	AVG	OBP	SLG
1914	Bos	A	5	10	1	2	0	2	.200	.200	.300
1915	Bos	A	42	92	16	29	4	21	.315	.376	.576
1916	Bos	A	67	136	18	37	3	15	.272	.322	.419
1917	Bos	A	52	123	14	40	2	12	.325	.385	.472
1918	Bos	A	95	317	50	95	11	66	.300	.410	.555
1919	Bos	A	130	432	103	139	29	114	.322	.456	.657
1920	NYY	A	142	458	158	172	54	137	.376	.530	.847
1921	NYY	A	152	540	177	204	59	171	.378	.512	.846
1922	NYY	A	110	406	94	128	35	99	.315	.434	.672
1923	NYY	A	152	522	151	205	41	131	.393	.545	.764
1924	NYY	A	153	529	143	200	46	121	.378	.513	.739
1925	NYY	A	98	359	61	104	25	66	.290	.393	.543
1926	NYY	A	152	495	139	184	47	146	.372	.516	.737
1927	NYY	A	151	540	158	192	60	164	.356	.487	.772
1928	NYY	A	154	536	163	173	54	142	.323	.461	.709
1929	NYY	A	135	499	121	172	46	154	.345	.430	.697
1930	NYY	A	145	518	150	186	49	153	.359	.493	.732
1931	NYY	A	145	534	149	199	46	163	.373	.495	.700
1932	NYY	A	133	457	120	156	41	137	.341	.489	.661
1933	NYY	A	137	459	97	138	34	103	.301	.442	.582
1934	NYY	A	125	365	78	105	22	84	.288	.447	.537
1935	Bos	N	28	72	13	13	6	12	.181	.359	.431
(22 seasons)			2503	8399	2174	2873	714	2213	.342	.474	.690

(a) Construct a dotplot of the season home run numbers for Babe Ruth.

(b) Comment on the basic shape of the distribution of home run numbers.

(c) Can you think of some possible reasons why Ruth had so many seasons with a small number of home runs?

(d) Since the number of at-bats is not constant across seasons, it may be better to consider the home run rate, which is obtained by dividing the number of home runs by the number of at-bats. So, for example, Ruth's home run rate in 1915 is given by

$$\text{home run rate} = \frac{4}{92} = .0435.$$

The home run rates for Ruth in the 22 seasons are given below:

0.000	0.043	0.022	0.016	0.035	0.067	0.118	0.109	0.086
0.079	0.087	0.070	0.095	0.111	0.101	0.092	0.095	0.086
0.090	0.074	0.060	0.083					

(e) Construct a dotplot of the home run rates.

(f) Summarize the key features of the home run rates. Explain why it is better to look at the home run rates instead of the home run numbers.

2.6. (Continuation of Exercise 2.5)   Consider again the home rate rates for Babe Ruth displayed in Exercise 2.5.

(a) Graph the home run rates with a time series plot using the grid below.

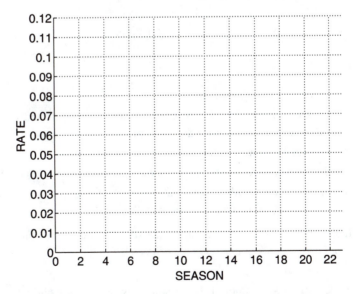

(b) What pattern do you see in the graph you made in (a)? Draw a smooth curve over the points.

(c) Using the smooth curve you constructed in (b), what was Babe Ruth's peak year with respect to home run rate? Is this the same year as the year when he hit the greatest number of home runs?

(d) Repeat parts (a)–(c) using Ruth's batting average statistic.

2.7. (Babe Ruth data from Exercise 2.5)

(a) Construct a dotplot of Ruth's season SLG values.

(b) Find the mean and median SLG values.

(c) Which average computed in (b) seems to be most representative of Ruth's slugging ability? Explain.

(d) In the table of Ruth's data, Ruth's career slugging percentage is .690. Is this the same as the mean SLG value that you computed in (b)? If the numbers are different, can you explain why?

2.8. (Babe Ruth data from Exercise 2.5) A stemplot of Ruth's season batting averages is shown below. (Here the stem is the first digit, so a batting average of .315 would have a stem of 3 and a leaf of 1.)

```
1 | 8
2 | 0
2 |
2 |
2 | 7
2 | 89
3 | 0011
3 | 222
3 | 4455
3 | 77777
3 | 9
```

(a) Compute a five-number summary.

(b) Looking back at the table of data in Exercise 2.3, find the years where Ruth's batting average was above the median.

(c) Find the years where Ruth's average was below the median.

(d) Based on (b) and (c), do you think the median is a representative AVG for Ruth?

(e) In the table of Ruth's data, Ruth's career AVG is equal to .342. Compare this value to the median and explain why they are different.

2.9. Table 2.9 shows the career pitching statistics for the Hall of Famer Sandy Koufax. (Koufax is the only pitcher to make ESPN's list of the top 50 Greatest Athletes of the 20th century.)

(a) Construct a stemplot of the season ERA values for Koufax (the breakpoint between the stem and the leaf will be at the decimal point).

(b) There are two clusters in this dataset. Identify the two clusters and explain which season years are identified with the two clusters.

TABLE 2.9
Pitching statistics for Sandy Koufax.

YR	TM	Lg	GP	W	L	PCT	IP	H	BB	SO	ERA
1955	Bro	N	12	2	2	.500	41.2	33	28	30	3.02
1956	Bro	N	16	2	4	.333	58.2	66	29	30	4.91
1957	Bro	N	34	5	4	.556	104.1	83	51	122	3.88
1958	LA	N	40	11	11	.500	158.2	132	105	131	4.48
1959	LA	N	35	8	6	.571	153.1	136	92	173	4.05
1960	LA	N	37	8	13	.381	175.0	133	100	197	3.91
1961	LA	N	42	18	13	.581	255.2	212	96	269	3.52
1962	LA	N	28	14	7	.667	184.1	134	57	216	2.54
1963	LA	N	40	25	5	.833	311.0	214	58	306	1.88
1964	LA	N	29	19	5	.792	223.0	154	53	223	1.74
1965	LA	N	43	26	8	.765	335.2	216	71	382	2.04
1966	LA	N	41	27	9	.750	323.0	241	77	317	1.73
(12 seasons)			397	165	87	.655	2324.1	1754	2396	2396	2.76

(c) Construct a stemplot of Koufax winning percentages (PCT). Describe the features of this dataset. Would it be accurate to say that Koufax was a successful pitcher? Why? What proportion of seasons did Koufax have a winning record?

2.10. (Sandy Koufax data from Exercise 2.9.)

(a) Draw a stemplot with five leaves per stem for the season strikeout counts. Describe the main features of this dataset. Are there any SO counts that appear unusually small or large? Is there any explanation for these unusual values?

(b) Find the mean and median SO number.

(c) Which average in (b) is the best measure of the center of the distribution? Explain. (It might help to look at the stemplot of the season strikeout counts from part (a).)

2.11. (Sandy Koufax data from Exercise 2.9)

(a) Compute the mean and median of Koufax season ERAs.

(b) Suppose Koufax played an additional season and his ERA was 6.50. Compute the mean and median of Koufax's ERAs with this additional data value.

(c) Which average (mean or median) changed the most when you added this new ERA value? Explain why this average changed.

(d) Koufax's career ERA was 2.76. Explain why this career ERA is different from the averages you computed in part (a).

2.12. Table 2.10 gives career pitching statistics for the Hall of Famer Bob Feller. Note that there are no statistics given for the three-year period 1942–1944—Feller served in the Navy during World War II between 1942 and 1945.

TABLE 2.10
Pitching statistics for Bob Feller.

YR	TM	Lg	GP	W	L	PCT	IP	H	BB	SO	ERA
1936	Cle	A	14	5	3	.625	62.0	52	47	76	3.34
1937	Cle	A	26	9	7	.563	148.2	116	106	150	3.39
1938	Cle	A	39	17	11	.607	277.2	225	208	240	4.08
1939	Cle	A	39	24	9	.727	296.2	227	142	246	2.85
1940	Cle	A	43	27	11	.711	320.1	245	118	261	2.61
1941	Cle	A	44	25	13	.658	343.0	284	194	260	3.15
1945	Cle	A	9	5	3	.625	72.0	50	35	59	2.50
1946	Cle	A	48	26	15	.634	371.1	277	153	348	2.18
1947	Cle	A	42	20	11	.645	299.0	230	127	196	2.68
1948	Cle	A	44	19	15	.559	280.1	255	116	164	3.56
1949	Cle	A	36	15	14	.517	211.0	198	84	108	3.75
1950	Cle	A	35	16	11	.593	247.0	230	103	119	3.43
1951	Cle	A	33	22	8	.733	249.2	239	95	111	3.50
1952	Cle	A	30	9	13	.409	191.2	219	83	81	4.74
1953	Cle	A	25	10	7	.588	175.2	163	60	60	3.59
1954	Cle	A	19	13	3	.813	140.0	127	39	59	3.09
1955	Cle	A	25	4	4	.500	83.0	71	31	25	3.47
1956	Cle	A	19	0	4	.000	58.0	63	23	18	4.97
(18 seasons)			570	266	162	.621	3827	3271	2581	2581	3.25

(a) Construct a stemplot of Feller's strikeout counts for his 18 seasons. What is the basic shape of this distribution? Are there any unusually small or large SO counts?

(b) It is difficult to tell which years Feller was especially good in striking out batters since the innings pitched (IP) is not constant across years. To adjust for the innings pitched, one can compute the strikeout ratio, the number of strikeouts divided by the innings pitched. Here are the strikeout ratios for Feller's 18 seasons:

1.23	1.01	0.87	0.83	0.82	0.76	0.82	0.94	0.66	0.59
0.51	0.48	0.45	0.42	0.34	0.42	0.30	0.31		

Graph these SO ratios against year number on the grid below. Describe any pattern that you see in the graph. What does that tell you about Feller's ability to strike out batters over time?

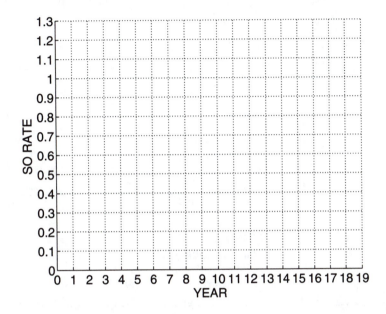

(c) Use Feller's lifetime average number of wins, losses, and strikeouts to fill in the missing years 1942–1944 and the partial year 1945. What are his new career totals in these categories?

2.13. (Bob Feller data from Exercise 2.12)

(a) Construct a dotplot of Feller's strikeout ratios for the 18 seasons. (These ratios are listed in part (b) of Exercise 2.12.)

(b) Compute the mean and median strikeout ratio. Explain why the mean is larger than the median for this dataset.

2.14. Figure 2.18 displays a histogram of the on-base percentages (OBP) for all 555 National League players who batted during the 1999 baseball season.

(a) Describe the basic shape of the histogram and any unusual features.

(b) What percentage of players had an OBP value between .3 and .4?

(c) What percentage of players had an OBP smaller than .2?

(d) There were six players that had an OBP value of 1 during the 1999 season. (This means that they reached base 100% of the time.) This may seem surprising—can you offer any possible explanation for these large values?

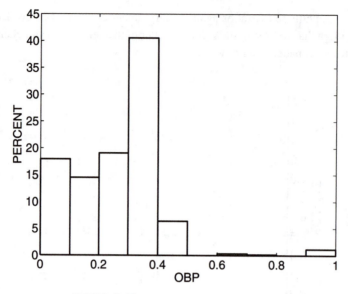

FIGURE 2.18
Histogram of OBPs of all 1999 NL players.

2.15. (Exercise 2.14 continued)   If one graphs the OBPs for only those National League hitters who had at least 300 at-bats (AB), one obtains the histogram in Figure 2.19. (In the following, we will refer to the players with at least 300 AB as the "regulars.")

   (a) Describe the shape of this histogram. Why does this histogram look so different from the histogram of the OBP of all NL players in Figure 2.18?

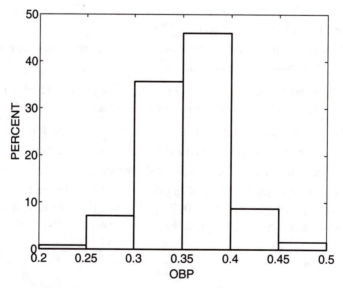

FIGURE 2.19
Histogram of OBPs for all 1999 NL players with at least 300 at-bats.

(b) What percent of NL regulars had an OBP value exceeding .3?

(c) What percent of NL regulars had an OBP value between .3 and .4?

(d) From the histogram, what OBP value would you consider outstanding? Why?

2.16. Figure 2.20 shows the number of home runs for the 1999 NL regular players with at least 300 at-bats.

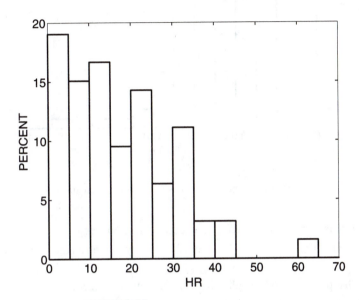

FIGURE 2.20
Home runs for all 1999 NL regular players.

(a) Describe the basic shape of the histogram and any unusual characteristics.

(b) What percentage of NL regular players hit fewer than ten home runs?

(c) What percentage of NL regular players hit more than 20 home runs?

(d) How many home runs did an NL regular player hit, on average?

(e) There were two hitters that hit at least 60 home runs in the 1999 season. Can you identify these players?

2.17. (Exercise 2.16 continued)   One explanation for the wide variation of home run totals shown in Figure 2.20 is that players had different numbers of at-bats. If we divide a player's home run count by his number of at-bats, one obtains a player's home run rate. (A rate of .08 means a batter hit a home run in 8% of his at-bats.) Figure 2.21 shows a histogram of the home run rates for the 1999 NL regulars.

(a) Describe the basic shape of this distribution of home run rates. Are there any unusual values?

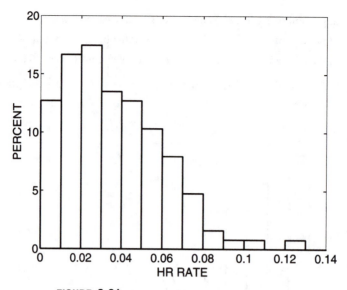

FIGURE 2.21
Histogram of home run rates for 1999 NL regulars.

(b) What is an average home run rate among these NL regulars?

(c) If a batter comes to bat 100 times, how many home runs do you expect him to hit?

(d) What percent of NL regulars had a home run rate under .02?

(e) A hitter is considered to be an unusually good home run hitter if his rate of hitting home runs exceeds 10% or .1. What percent of regular NL hitters fall into this classification? Can you guess which players are in this class?

2.18. Table 2.11 displays the length (in minutes) of a sample of 58 baseball games in 1999.

TABLE 2.11
Length in minutes of a selection of 1999 baseball games.

158	208	181	145	146	164	143	176	256	193
136	179	133	151	209	170	158	265	151	186
152	156	172	140	150	170	240	160	165	179
175	193	158	168	165	157	152	241	198	177
176	187	184	265	187	130	159	167	259	172
169	192	203	214	186	211	175	171		

(a) Construct a stemplot of these game times.

(b) What is the general shape of the distribution of times? Are there any unusually short or long games?

(c) What is a median length of a baseball game?

(d) What fraction of games are over three hours?

(e) Why is there such a large spread of times? Can you think of some variables that might influence the length of a baseball game?

2.19. (Length of baseball games from Exercise 2.18)   A stemplot of the times of games (in minutes) is shown below.

```
13 | 036
14 | 0356
15 | 01122678889
16 | 0455789
17 | 001225566799
18 | 146677
19 | 2338
20 | 389
21 | 14
22 |
23 |
24 | 01
25 | 69
26 | 55
```

(a) Find the five-number summary of the lengths of the 1999 baseball games.

(b) Approximately the middle half of the game lengths is between ____ and ____ .

(c) The mean and standard deviation of the game lengths are 179 minutes and 32.5 minutes, respectively. Find an interval defined by

$$(\text{mean} - \text{standard deviation}, \quad \text{mean} + \text{standard deviation}).$$

(d) Find the proportion of games that fall in the interval you found in part (c).

(e) If the data is normally distributed, then one would expect 68% of the data to fall in the interval in (c). Here you should find that the proportion in (d) is significantly larger than .68. Can you explain why?

2.20. For each of 58 baseball games in 1999, the total number of home runs was recorded—the numbers are shown in Table 2.12.

TABLE 2.12
Number of home runs hit in a sample of 1999 baseball games.

2	2	1	2	0	4	2	1	1	2	2	5
0	2	2	2	1	1	0	3	0	0	2	2
0	2	4	0	1	3	2	1	2	0	2	7
2	5	2	3	0	6	5	1	5	2	3	0
3	1	5	3	5	3	0	4	4	3		

(a) Construct a frequency table for the number of home runs.

(b) Graph the frequencies using a bar chart.

(c) Find the proportion of games where exactly two home runs are hit.

(d) Find the proportion of games where four or more home runs are hit.

(e) What is an average number of home runs hit during a game? Explain how you computed this average number.

2.21. Table 2.13 displays basic batting statistics for the 1999 American League teams.

TABLE 2.13

Batting statistics for the 1999 American League teams.

City	Avg	OBP	SLG	AB	R	H	2B	3B	HR	RBI
Anaheim	0.256	0.322	0.395	5494	711	1404	248	22	158	673
Baltimore	0.279	0.353	0.447	5637	851	1572	299	21	203	804
Boston	0.278	0.350	0.448	5579	836	1551	334	42	176	808
Chicago	0.277	0.337	0.429	5644	777	1563	298	37	162	742
Cleveland	0.289	0.373	0.467	5634	1009	1629	309	32	209	960
Detroit	0.261	0.326	0.443	5481	747	1433	289	34	212	704
Kansas City	0.282	0.348	0.433	5624	856	1584	294	52	151	800
Minnesota	0.264	0.328	0.384	5495	686	1450	285	30	105	643
New York	0.282	0.366	0.453	5568	900	1568	302	36	193	855
Oakland	0.259	0.355	0.446	5519	893	1430	287	20	235	845
Seattle	0.269	0.343	0.455	5572	859	1499	263	21	244	825
Tampa Bay	0.274	0.343	0.411	5586	772	1531	272	29	145	728
Texas	0.293	0.361	0.479	5651	945	1653	304	29	230	897
Toronto	0.280	0.352	0.457	5642	883	1580	337	14	212	856

(a) Draw a stemplot of the batting averages (AVG) for the 14 AL teams.

(b) Find the median batting average and find a team that has a batting average close to the median value.

(c) Find the team that has the highest batting average and the team with the lowest batting average.

(d) For a second statistic of your choice, construct a stemplot. Find a team that has an average value of the statistic. Also find a team that has the smallest value, and a team that has the largest value of the statistic that you chose.

2.22. (Batting statistics for 1999 AL teams from Exercise 2.21)

(a) Draw a stemplot of the HR numbers for the 14 AL teams.

(b) Find the five-number summary.

(c) Approximately _____ percent of the HR numbers are smaller than 198.

(d) Approximately _____ percent of the HR numbers are larger than 158.

(e) Find a team that has an HR number that is close to the average.

2.23. Table 2.14 gives the year of birth and earned run average (ERA) for all 57 pitchers who have been inducted into the Baseball Hall of Fame.

TABLE 2.14
Year of birth and ERA for pitchers who have been selected to the Baseball Hall of Fame.

Pitcher	Birthyear	ERA	Pitcher	Birthyear	ERA
Grover Cleveland Alexander	1887	2.56	Sandy Koufax	1935	2.76
Charles "Chief" Bender	1884	2.46	Bob Lemon	1920	3.23
Mordecai "Three Finger" Brown	1876	2.06	Juan Marichal	1937	2.89
Jim Bunning	1931	3.27	Richard "Rube" Marquard	1889	3.08
Steve Carlton	1944	3.22	Christy Mathewson	1880	2.13
Jack Chesbro	1874	2.68	Joe McGinnity	1871	2.66
John Clarkson	1861	2.81	Hal Newhouser	1921	3.06
Stan Coveleski	1889	2.89	Charles "Kid" Nichols	1869	2.95
W.A. "Candy" Cummings	1848	2.78	Phil Niekro	1939	3.35
Jay Hanna "Dizzy" Dean	1911	3.02	Satchel Paige	1906	3.29
Don Drysdale	1936	2.95	Jim Palmer	1945	2.86
Urban "Red" Faber	1888	3.15	Herb Pennock	1894	3.60
Bob Feller	1918	3.25	Gaylord Perry	1938	3.11
Rollie Fingers	1946	2.90	Eddie Plank	1875	2.35
Edward "Whitey" Ford	1928	2.75	Eppa Rixey	1891	3.15
Rube Foster	1888	2.36	Robin Roberts	1926	3.41
James "Pud" Galvin	1856	2.87	Amos Rusie	1871	3.07
Bob Gibson	1935	2.91	Nolan Ryan	1947	3.17
Vernon "Lefty" Gomez	1908	3.34	Tom Seaver	1944	2.86
Burleigh Grimes	1893	3.53	Warren Spahn	1921	3.09
Robert "Lefty" Grove	1900	3.06	Don Sutton	1945	3.26
Jesse "Pop" Haines	1878	3.64	A.C. "Dazzy" Vance	1891	3.24
Waite Hoyt	1899	3.59	George "Rube" Waddell	1876	2.16
Carl Hubbell	1903	2.98	Ed Walsh	1881	1.82
Jim "Catfish" Hunter	1946	3.26	Mickey Welch	1859	2.71
Ferguson Jenkins	1943	3.34	Hoyt Wilhelm	1923	2.52
Walter Johnson	1887	2.16	Vic Willis	1876	2.63
Addie Joss	1880	1.89	Early Wynn	1920	3.54
Tim Keefe	1857	2.62	Denton "Cy" Young	1867	2.63

(a) Draw a stemplot of the ERAs of these Hall of Fame pitchers.

(b) Discuss the general shape of the distribution of ERAs and any unusual characteristics of this data.

(c) Find an average ERA and a Hall of Fame pitcher who has (approximately) this average ERA.

(d) Find the pitchers who have the lowest and highest ERAs.

(e) What might explain this wide range of ERAs if all of these pitchers are in the Hall of Fame?

2.24. (Exercise 2.23 continued)   Consider the years of birth of the pitchers in the Hall of Fame.

    (a) Construct a frequency table of the years of birth, using the intervals in the table below. Place the counts and the proportions in the table.

Interval for year of birth	Count	Proportion
1840 to 1859		
1860 to 1879		
1880 to 1899		
1900 to 1919		
1920 to 1939		
1940 to 1959		

    (b) Construct a histogram from the above frequency table.

    (c) Describe the basic shape of the distribution.

    (d) You should note that the Hall of Fame pitchers are not uniformly distributed over the six ERAs 1840–1859, 1860–1879, ..., 1940–1959. Can you explain this pattern? Were pitchers better during particular years of baseball?

2.25. The salaries of the players on the 1999 Atlanta Braves are displayed in Table 2.15.

TABLE 2.15

Salaries of the 1999 Atlanta Braves.

Greg Maddux	$10,600,000	Russ Springer	$950,000
Andres Galarraga	$8,250,000	Rudy Seanez	$625,000
John Smoltz	$7,750,000	Eddie Perez	$550,000
Tom Glavine	$7,000,000	Ozzie Guillen	$500,000
Javy Lopez	$5,250,000	Andruw Jones	$330,000
Mark Wohlers	$5,200,000	Brian Hunter	$300,000
Ryan Klesko	$4,750,000	Kerry Lightenberg	$255,000
Brian Jordan	$4,600,000	Kevin Millwood	$230,000
Chipper Jones	$4,000,000	Mike Cather	$225,000
Walt Weiss	$3,000,000	John Rocker	$217,500
Bret Boone	$2,900,000	Randall Simon	$202,500
Otis Nixon	$1,500,000	Wes Helms	$200,000
Gerald Williams	$1,400,000	Kevin McGlinchy	$200,000
Mike Remlinger	$1,100,000	Damian Moss	$200,000
Keith Lockhart	$1,000,000	Odalis Perez	$200,000

    (a) Construct a stemplot of the salaries.

    (b) Describe the basic shape of the data. Are there any unusually small or large salaries?

    (c) Find the median salary and a player who has this median salary.

    (d) Can you explain why there are no salaries below $200,000?

2.26. (Exercise 2.25 continued)   Table 2.16 displays the salaries of the 1999 Florida Marlins.

TABLE 2.16
Salaries of the 1999 Florida Marlins.

Alex Fernandez	$7,000,000	Kirt Ojala	$240,000
Cliff Floyd	$2,500,000	Kevin Orie	$240,000
Jorge Fabregas	$1,850,000	Dave Berg	$235,000
Livan Hernandez	$1,575,000	Luis Castillo	$225,000
Matt Mantei	$ 735,000	Todd Dunwoody	$222,000
Mark Kotsay	$ 280,000	Mike Redmond	$220,000
Dennis Springer	$ 275,000	Alex Gonzalez	$201,000
Vic Darensbourg	$ 265,000	Braden Looper	$201,000
Brian Meadows	$ 265,000	Mike Lowell	$201,000
Jesus Sanchez	$ 265,000	Preston Wilson	$201,000
Derrek Lee	$ 255,000	Bruce Aven	$200,000
Antonio Alfonseca	$ 250,000	Guillermo Garcia	$200,000
Craig Counsell	$ 245,000	Tim Hyers	$200,000
Brian Edmondson	$ 240,000		

(a) Answer questions (a), (b), (c) of Exercise 2.25 for this salary dataset.

(b) Compare the distributions of salaries of the 1999 Braves with the 1999 Marlins.

2.27. (Florida Marlins salary data from Exercise 2.26)   A histogram of the salaries is shown below.

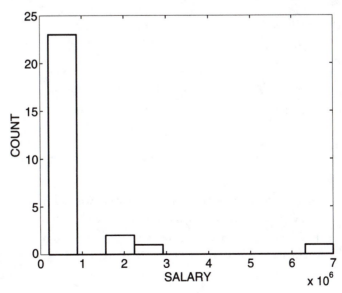

48 TEACHING STATISTICS USING BASEBALL

(a) Based on the shape of the distribution of these salaries, do you expect the mean to be larger than, equal to, or smaller than the median?

(b) Compute the mean and median of the salaries of the 1999 Florida Marlins.

(c) Do the values you computed in (b) agree with your expectation in (a)?

2.28. In baseball, there are four famous (or notable) years with respect to home runs: 1927 when Babe Ruth hit 60 home runs, 1961 when Roger Maris hit 61, 1998 when Mark McGwire hit 70 home runs, and 2001 when Barry Bonds hit 73 home runs. For the four years 1927, 1961, 1998, and 2001, the number of home runs was collected for each player who had at least 400 at-bats. A player's home run count was classified into one of the six groupings, 0, 1–5, 6–10, 11–20, 21–40, 41 and more. The frequency table for the home runs for each of the three years is presented in Table 2.17.

TABLE 2.17
Classification of home run numbers for all players with at least 400 AB for the years 1927, 1961, 1998, 2001.

1927 Players

Number of home runs	0	1–5	6–10	11–20	21–40	41 and more
Count	7	50	19	12	3	2
Proportion						

1961 Players

Number of home runs	0	1–5	6–10	11–20	21–40	41 and more
Count	0	17	13	30	27	7
Proportion						

1998 Players

Number of home runs	0	1–5	6–10	11–20	21–40	41 and more
Count	0	20	30	53	59	12
Proportion						

2001 Players

Number of home runs	0	1–5	6–10	11–20	21–40	41 and more
Count	0	10	16	37	112	12
Proportion						

(a) For each table, find the proportion of players who fall in each of the six categories. Place the proportions in the rows labeled "Proportion."

(b) In 1927, what proportion of players hit more than 20 home runs? Repeat this computation for the 1961, 1998, and 2001 years. Place your answers in the table below.

Year	Proportion hitting more than 20 home runs
1927	
1961	
1998	
2001	

(c) Based on your computations in (a) and (b), would you say that in some years it was more difficult to hit a home run than in other years? Explain.

2.29. Table 2.18 gives average ballpark attendance for each of the major league teams in 1999. The "home avg" column gives the average attendance for all games played at home, and the "away avg" column gives the average attendance for all games played on the road.

TABLE 2.18
Average attendance figures for all ML teams in 1999.

American League			National League		
Team	home avg	away avg	Team	home avg	away avg
Anaheim	27,806	27,703	Arizona	36,852	30,203
Baltimore	42,880	27,658	Atlanta	40,554	31,027
Boston	30,201	30,165	Chicago	36,074	34,574
Chicago	17,616	28,821	Cincinnati	25,136	28,392
Cleveland	42,845	30,311	Colorado	42,487	27,990
Detroit	25,017	27,041	Florida	17,556	26,602
Kansas City	19,247	27,566	Houston	33,407	26,033
Minnesota	14,849	28,068	Los Angeles	38,303	32,363
New York	40,649	33,574	Milwaukee	21,817	25,139
Oakland	17,711	28,100	Montreal	9,546	28,518
Seattle	35,998	30,588	New York	34,505	31,712
Tampa Bay	21,599	27,591	Philadelphia	22,816	27,767
Texas	34,681	26,002	Pittsburgh	20,475	27,471
Toronto	26,709	26,998	San Diego	31,154	30,491
			San Francisco	25,655	30,381
			St. Louis	40,963	35,700

(a) Construct a dotplot of the home attendance averages for all AL teams. Comment on the basic shape of the data and note any unusual values.

(b) Repeat (a) for the home averages for the NL teams.

(c) Find the AL team that has the smallest average home attendance, the team with the largest average, and a team that has a median average.

(d) Repeat (c) for the NL teams.

(e) Can you explain the great variation in home attendance averages? Do you think the variation can be explained solely by the differences in the sizes of the cities?

2.30. (Ballpark attendance data from Exercise 2.29)

(a) Find the five-number summary of the home average attendances for the AL teams.

(b) Find the five-number summary of the home average attendances for the NL teams.

(c) By comparing the medians you found in (a) and (b), which league tends to draw more fans at their home games?

(d) Which league seems to have more variation in home attendance among its teams? Which number should you be computing to measure the spread of the AL and NL datasets?

2.31. (Ballpark attendance data from Exercise 2.29)

(a) Draw a stemplot of the home average attendances for all 30 Major League teams.

(b) Draw a stemplot of the away average attendances for all 30 Major League teams.

(c) Find the five-number summaries of the home average attendances and the away average attendances.

(d) Compare the medians of the home and away average attendances.

(e) Compare the spreads of the home and away average attendances. Explain why one dataset has a much smaller spread than the other dataset.

2.32. Table 2.19 gives the batting side (L = left, R = right, B = both sides) and the throwing hand (L = left, R = right) for 40 randomly selected ballplayers who were born in 1950 or later.

TABLE 2.19
Batting side and throwing hand for forty randomly selected players.

Batting Side																		
L	R	L	B	R	R	L	R	R	B	R	R	R	L	R	L	R	L
R	L	B	R	L	R	R	R	R	L	R	R	R	L	R	R	L	R
R	B	L	R														

Throwing Hand																		
R	R	R	R	R	R	R	L	L	R	R	R	L	R	R	R	R
R	R	L	R	R	R	R	R	R	R	R	R	L	R	R	L	L	R
R	R	R	R														

(a) Construct a frequency table for the batting side of the 40 players.

(b) Graph the proportions of left, right, and both sides batters using a bar chart.

(c) Repeat (a) and (b) for the throwing hand of the 40 players.

(d) Do you think that the proportion of left-handed batters is less than, equal to, or greater than the proportion of lefties in the American population (which is about 10%)? Explain.

(e) Do you think that the proportion of left-handed throwers is less than, equal to, or greater than the proportion of lefties in the American population? Explain.

2.33. Figure 2.22 shows the location of the home runs and batted balls that landed in the outfield for Mark McGuire during the 1999 baseball season.

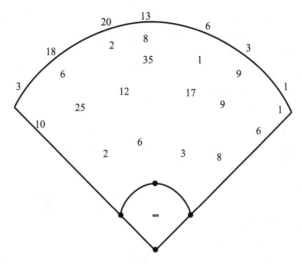

FIGURE 2.22
Results of all of Mark McGwire's plate appearances during the 1999 season.

(a) In this diagram, home runs are indicated by the numbers above the outfield grass. According to this diagram, how many home runs did McGwire have during the 1999 season?

(b) McGwire's home runs can be classified into those falling in (over) the left field stands, those in center field, and those in right field. Find the number and proportion of home runs of the three locations. Put your answers in the table below.

McGwire's home runs

Location	Count	Proportion
Left		
Center		
Right		

(c) In a similar fashion, classify all of McGwire's batted balls that landed in the outfield.

McGwire's outfield balls

Location	Count	Proportion
Left		
Center		
Right		

(d) Based on your work in (a) and (b), would it be reasonable to say that McGwire was a so-called "pull hitter" (that is, a right-handed hitter who tends to hit the ball to left field)? Explain.

2.34. Table 2.20 gives the number of hits for Sammy Sosa for each of the games in the 1999 season where he had exactly four at-bats.

TABLE 2.20
Number of hits for Sammy Sosa for all games in 1999 where he had exactly four at bats.

0	0	1	1	4	1	1	0	0	3	0	2
1	1	1	0	2	0	3	0	3	1	1	1
1	1	2	1	1	0	2	1	0	2	2	2
2	1	0	2	2	0	1	1	1	3	0	1
1	1	2	0	1	0	1	2	1	0	3	2
2	0	1	2	0	2	1	1	1	2	0	2
0	2	1	0	1							

(a) Compute a frequency table of hits; put the counts and proportions in the table below.

Hits	Count	Proportion

(b) Construct a histogram of the hit numbers.

(c) Find the proportion of *these* games where Sosa got at least one hit.

2.35. (Hits of Sammy Sosa for games during the 1999 season from Exercise 2.34)

(a) Find the mean number of hits.

(b) Find the median number of hits.

(c) Explain why the mean and median are different for these data.

2.36. Table 2.21 gives the number of hits for Ken Griffey, Jr. for each of the games in the 1999 season where he had exactly four at-bats.

TABLE 2.21
Number of hits by Ken Griffey, Jr. for all 1999 games where he had exactly four at bats.

4	1	1	1	0	0	0	1	0	2	0	1
2	1	2	3	2	2	2	1	2	1	0	1
0	1	1	3	1	0	0	1	1	2	2	0
2	1	0	1	2	1	1	1	1	2	1	1
2	2	2	1	0	1	1	0	1	1	1	0
2	0	0	1	0	2	2	0	1	0	2	0

(a) Construct a stemplot of the number of hits.

(b) Find the proportion of *these* games where Griffey was hitless.

(c) If you were watching Griffey play and he had exactly four at-bats, what would be the most likely number of hits?

2.37. Table 2.22 below gives the date, innings pitched, and number of earned runs allowed for each game that Charles Nagy pitched during the 1999 baseball season.

TABLE 2.22
Pitching statistics for Charles Nagy for each pitched game during 1999 season. Under innings pitched (IP), the notation "5.1" means $5\frac{1}{3}$ innings, and "5.2" means $5\frac{2}{3}$ innings.

Game	Date	IP	ER	Game	Date	IP	ER
1	04/09/99	7.0	3	18	07/17/99	4.0	4
2	04/17/99	7.2	1	19	07/22/99	6.0	4
3	04/22/99	6.2	4	20	07/27/99	8.0	5
4	04/27/99	6.2	3	21	08/01/99	8.0	5
5	05/02/99	3.1	8	22	08/07/99	5.0	7
6	05/08/99	3.2	5	23	08/13/99	7.0	1
7	05/14/99	6.2	2	24	08/18/99	6.0	5
8	05/19/99	8.0	1	25	08/23/99	8.0	4
9	05/25/99	7.0	1	26	08/28/99	8.0	0
10	05/31/99	6.0	1	27	09/02/99	7.0	1
11	06/06/99	6.0	2	28	09/07/99	6.0	2
12	06/12/99	6.1	3	29	09/13/99	5.2	5
13	06/18/99	5.2	3	30	09/18/99	6.2	3
14	06/23/99	7.0	4	31	09/23/99	5.0	6
15	06/28/99	7.0	1	32	09/29/99	7.0	5
16	07/03/99	2.0	8	33	10/03/99	1.0	3
17	07/08/99	7.0	1				

(a) Construct a stemplot of the innings pitched.

(b) Comment on the basic features of this dataset.

(c) Based on the display, what was a median number of innings pitched for Nagy this season?

(d) Construct a stemplot of number of earned runs for the 33 games pitched.

(e) What is the shape of the stemplot of earned runs?

(f) Find the proportion of games where Nagy allowed one earned run or fewer.

2.38. The stemplot below displays the batting averages for all 1999 players in both leagues who had at least 400 at bats. (Pitchers are not included in this dataset.)

```
19 | 5
20 |
21 |
22 | 6
23 | 16
24 | 00145557
25 | 01122333555668899
26 | 0333456667788899
27 | 01244555555677777788899
28 | 00000011112234455555578889
29 | 0000112333334444455556778888
30 | 0001111123334444444567899
31 | 01233455569
32 | 00234556668
33 | 22335678
34 | 29
35 | 7
36 |
37 | 9
```

(a) Describe the general shape of this data and any unusual characteristics.

(b) The mean and standard deviation of these batting averages are .287 and .027, respectively. Find an interval where you expect 95% of the AVGs to fall.

(c) Find the actual proportion of AVGs that fall in the interval you found in part (b).

2.39. The stemplot below displays the number of home runs allowed per nine innings (HR9 = $9 \times$ HR/IP) for all 1999 pitchers who pitched at least 100 innings. (Here the breakpoint is to the right of the tenth's place, so a stem of 14 and a leaf of 2 would record an HR9 value of 1.42.)

3	7
4	5
5	0278
6	0355667799
7	003345999
8	001336888999
9	02333445567788899
10	00001334455567799
11	11223344555569
12	0122233444466789
13	0023667888
14	001234777899
15	0111334569
16	023378
17	4
18	357
19	2234
20	
21	
22	0

(a) Describe the general shape of the data and note any unusual characteristics. Are there pitchers who allowed an unusually high or low number of home runs?

(b) Compute the five-number summary.

(c) Of the 148 pitchers, what proportion allowed less than one home run for nine innings?

(d) What proportion of pitchers allowed more than two home runs for nine innings?

(e) Suppose that your favorite pitcher allows more than two home runs each nine innings. Based on your work in (a)–(c), how would you rate this performance?

Further Reading

Devore and Peck (2000) and Moore and McCabe (1998) provide good descriptions of the exploratory methods used in this chapter to describe a single batch of data. Chapter 2 of Albert and Bennett (2001) illustrates the use of data analysis methods on baseball data.

3

Comparing Batches and Standardization

What's On-Deck?

In this chapter, we illustrate some basic data analysis tools for comparing two or more datasets. It is popular among baseball fans to make comparisons between individual players. In Case Study 3-1, we compare two of the most popular sluggers, Barry Bonds and Ken Griffey Jr. By using parallel boxplots and time series plots, we compare the batting performance of the two players. In Case Study 3-2, we compare Robin Roberts and Whitey Ford, two Hall of Fame pitchers who pitched in the opening game of the 1950 World Series. When we think of great individual home run accomplishments, we naturally think of Babe Ruth, who hit 60 home runs in 1927, Roger Maris, who hit 61 in 1961, Mark McGwire, who hit 70 in 1998, and Barry Bonds, who hit 73 home runs in 2001. To better understand these hitting accomplishments, one should look at the general pattern of home run hitting during these four seasons, and Case Study 3-3 compares the team home run rates for these seasons. The next study looks at the slugging percentages of all players in the 1999 season that had at least 400 at-bats. We will see that the distribution of slugging percentages has a distinctive bell shape and this pattern makes it easy to find intervals that contain a given percentage of the data. We conclude the chapter in Case Study 3-5 by looking at four of the highest season batting averages in recent baseball history. One can assess the greatness of each hitting accomplishment by looking at each batting average in the context of all batting averages for that particular season. By computing standardized scores, we can compare these hitting accomplishments and say which player had the best average relative to his peers.

Case Study 3-1: Barry Bonds and Junior Griffey

Topics Covered: Stemplot, five-number summary, time series plot, boxplot   In this first case study, we statistically compare Ken Griffey Jr. (Junior) and Barry Bonds (Barry), two of the best current ballplayers.

Let's make some initial comparisons:

1. Both players have famous fathers. Junior's dad is Ken Griffey, who played right field for the Reds in the 1970s. Barry's dad is Bobby Bonds, who played center field for the Giants in the 1970s. Bonds Sr. was noted for his combination of speed and power, which was reflected in a large number of stolen bases and home runs.

2. Junior was born in Ohio, went to the west coast to play for Seattle, and is now back in Ohio playing for the Reds. Barry was born in California, initially played in the east for the Pirates, and is now playing in California for the Giants.

3. Ages: In the 2002 baseball season, Junior was 32 and Barry was 38, so Barry is the more senior player. Junior started in the major leagues at age 19 and Barry began playing major league ball at 22.

4. Both players bat left-handed.

5. Barry has won five MVP (Most Valuable Player) titles and Junior has won one MVP title.

Who is the better hitter? Before we compare Barry and Junior, some general truths about evaluating hitting should be stated. The objective for a baseball team is to score **runs**.

1. How does a team score runs?

   - batters get **on base**
   - other batters **advance** the runners to home by hits, walks, HBP, or errors

2. A measure of a hitter's ability to get on-base is the on-base percentage (OBP):

$$\text{OBP} = \frac{(\text{H} + \text{BB} + \text{HBP})}{(\text{AB} + \text{BB} + \text{HBP} + \text{SF})}.$$

The total number of plate appearances is $\text{PA} = \text{AB} + \text{BB} + \text{HBP} + \text{SF}$ and the number of times on-base is $\text{H} + \text{BB} + \text{HBP}$. So OBP is the fraction of PAs that are on-base.

3. A measure of a hitter's ability to advance runners is the slugging percentage (SLG):

$$\text{SLG} = \text{TB}/\text{AB}$$

Here TB is total bases = number of singles +2 (number of doubles) +3 (number of triples) +4 (number of home runs)

4. As discussed in Chapter 1, a simple measure that combines on base ability with advancement ability is the OPS statistic:

$$\text{OPS} = \text{OBP} + \text{SLG}.$$

For example, Barry in 2001 had an OBP of .515 and an SLG of .863, so his 2001 OPS was OPS = .515 + .863 = 1.378.

Comparing OPS. Table 3.1 gives the season OPS values for Barry and Junior for their careers (through the 2002 season).

TABLE 3.1
Career hitting statistics for Barry Bonds and Ken Griffey, Jr.

	Barry				Junior			
Year	Age	OBP	SLG	OPS	Age	OBP	SLG	OPS
1986	22	0.33	0.416	0.746				
1987	23	0.329	0.492	0.821				
1988	24	0.368	0.491	0.859				
1989	25	0.351	0.426	0.777	19	0.329	0.42	0.749
1990	26	0.406	0.565	0.971	20	0.366	0.481	0.847
1991	27	0.41	0.514	0.924	21	0.399	0.527	0.926
1992	28	0.456	0.624	1.08	22	0.361	0.535	0.896
1993	29	0.458	0.677	1.135	23	0.408	0.617	1.025
1994	30	0.426	0.647	1.073	24	0.402	0.674	1.076
1995	31	0.431	0.577	1.008	25	0.379	0.481	0.86
1996	32	0.461	0.615	1.076	26	0.392	0.628	1.02
1997	33	0.446	0.585	1.031	27	0.382	0.646	1.028
1998	34	0.438	0.609	1.047	28	0.365	0.611	0.976
1999	35	0.389	0.617	1.006	29	0.384	0.576	0.96
2000	36	0.45	0.707	1.157	30	0.385	0.549	0.934
2001	37	0.515	0.863	1.378	31	0.365	0.533	0.898
2002	38	0.582	0.799	1.381	32	0.358	0.426	0.784

We use back-to-back stemplots in Figure 3.1 to make a comparison—we break an OPS value like 1.157 into a stem of 11 and a leaf of 57. We use one-digit leaves, so the OPS value of 1.157 is represented by a 5 leaf on the 11 stem line.

```
         BARRY OPS    JUNIOR OPS

                4  | 7  | 4
                7  | 7  | 8
                2  | 8  | 4
                5  | 8  | 699
                2  | 9  | 23
                7  | 9  | 67
             4300  | 10 | 222
              877  | 10 | 7
                3  | 11 |
                5  | 11 |
                   | 12 |
                   | 12 |
                   | 13 |
               87  | 13 |
```

FIGURE 3.1
Back-to-back stemplots for Barry Bonds and Ken Griffey Jr. season OPS values.

We summarize each dataset by a *five-number summary*—this is

$$(LO, QL, M, QU, HI)$$

where LO, HI are the lowest and highest values in the dataset, QL, QU are the lower and upper quartiles, and M is the median.

We illustrate these calculations for Barry's data. Here there are $n = 17$ values. The position of the median is $pos(M) = (17 + 1)/2 = 9$. The median is the 9th value, which is 1.031. To find the quartiles, we drop the middle observation and divide the 16 remaining observations into two groups of 8, and find the median of the lower 8 and the median of the upper 8. We get

$$QL = .892, \quad QU = 1.108.$$

The smallest value is .746 and the largest is 1.381.

So Barry's OPS values five-number summary is (.746, .892, 1.031, 1.108, 1.381). In a similar fashion, we compute Junior's OPS values 5-number summary: (.749, .860, .930, 1.020, 1.076).

Essentially, a five-number summary divides the data into quarters. For Barry's data, we can say that (approximately) 25% of the data falls between .746 and .892, 25% falls between .892 and 1.031, and so on.

Before we graph these two datasets, we look for possible outliers. Using a standard rule, we say that an extreme observation is worthy of special attention if it falls outside one step from the lower and upper quartiles, where a step is defined to be

$$step = 1.5 \times (QU - QL).$$

To illustrate for Barry's data,

- step $= 1.5 \times (1.108 - .892) = .324$

- Outliers are observations that have values smaller than

$$QL - step = .892 - .324 = .568$$

and larger than

$$QU + step = 1.108 + .324 = 1.432.$$

- Looking back at the stemplot of Barry's OPS values, we see that there no outliers in this dataset. However, most baseball fans would say that Bonds' OPS values in the 2001 and 2002 seasons are unusually high—in fact, these season OPS are among the largest season OPS of all time.

A boxplot is a graph of a five-number summary. One draws this by

- drawing a number line covering all of the values,

- drawing a box, where the ends of the box correspond to the quartiles QL and QU and a vertical line is drawn through the box at the median M,

- drawing lines (whiskers) out to the most extreme values that are not considered outliers,

- indicating outliers by plotting a special symbol.

Side-by-side boxplots are most useful in comparing datasets. In Figure 3.2 we show boxplots for the Griffey data and the Bonds data on the same scale. (We indicate Bonds' 2001 and 2002 values as outliers although they are not formal outliers using our rule.)

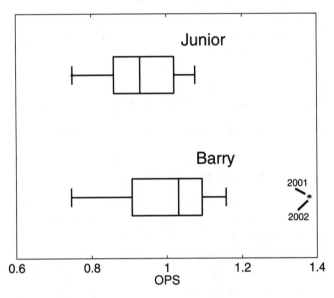

FIGURE 3.2
Parallel boxplots of the OPS values for Barry Bonds and Ken Griffey Jr.

What have we learned in this comparison?

- There is a general tendency for Barry to hit better (using the OPS statistic) than Griffey.

- On the average, Barry hits .101 higher than Junior on the OPS scale. We are comparing medians here:

$$\text{Median(Barry)} - \text{Median(Junior)} = 1.031 - .930 = .101$$

Adjusting Comparison for Ages. This comparison may be viewed as unfair, since Junior is six years younger than Barry and has more baseball seasons left. Maybe Barry and Junior may be perceived more equal after both have finished their careers.

Let's consider a different type of comparison that accounts for the ages of the two players.

What does it mean for Player A to have more ability than Player B? A popular way to model the relationship between ability and age is with the curve shown in Figure 3.3. In this curve, ability is low at the start of one's career, grows until mid-career, and then declines with advancing years. The peak ability, age at peak, the steepness of incline and decline, would be expected to vary from player to player.

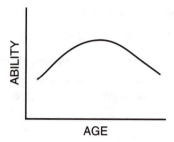

FIGURE 3.3
Expected ability curve for a Major League player.

Looking at the graph, we can distinguish between two measures of ability.

- *peak ability*—this is the player's ability at the top of the curve,

- *career ability*—this is the total accumulation of a player's ability over time. Think of the area under the above curve as representing career performance.

So when you say Barry is "better than" Junior, you should be clear what you're talking about. One player might play better at his peak, and the second player might be better in his total performance over his Major League career.

We can look at Barry and Junior's hitting performance by plotting the season OPS against age. Here is the data (Junior started playing at 19 and Barry at 22) and Figure 3.4 graphs the OPS against age for both players.

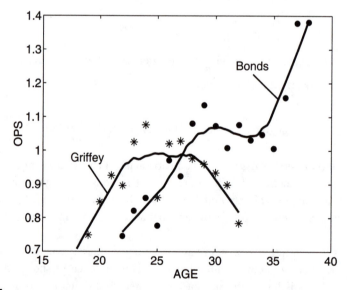

FIGURE 3.4
Time series plots of Barry Bonds' and Ken Griffey Jr.'s OPS values. Smooth curves are drawn through the points to show the basic patterns.

Drawing separate smooth curves over the points to show the basic patterns, what do we see?

- Junior's OPS values steadily increased until age 24, but they have consistently fallen off until the current year (2002). This is a bit surprising since most people peak around 28–30. His relatively low OPS values in the years 2000–2002 could be explained by the transition to the National League and his injuries.

- Barry's OPS values also increased steadily until about age 29 and he remained at a high performance level in his late 30's.

- Based on this graph, Junior reached his peak at a younger age than Barry, his peak is lower, and his performance deteriorated more rapidly after achieving its peak.

Case Study 3-2: Robin Roberts and Whitey Ford

Topics Covered: Stemplot, time series plot   In this case study, we discuss two great pitchers, Robin Roberts and Whitey Ford. In Phillies history, there were three great seasons:

1. 1950, when the Whiz Kids won the NL pennant—they were a young team that won the pennant on the last game of the season. (They lost to the Yankees in the World Series in four games.)
2. 1980, when they won the World Series against the Royals.
3. 1993, when they won the NL pennant and lost in six games to Toronto. (Joe Carter, who hit the series-ending home run, comes to mind.)

We compare two pitchers, both in the Hall of Fame, who played in the 1950 World Series.

Robin Roberts, born September 30, 1926, was a great pitcher for the 1950 Whiz Kids. He pitched for Philadelphia between 1948 and 1961. He won 286 games—many seasons he won over 20 games. His pitching was notable for his great control. He's still alive and recently wrote a book on the Whiz Kids.

We will compare Roberts with a great Yankees pitcher, Whitey Ford. Ford, born October 21, 1928, is notable for having 10 World Series wins. He won the Cy Young award in 1961 with a record of 25–4. (He was overshadowed that year by Roger Maris who hit 61 home runs.) His career win/loss record was 236–106. Like Roberts, Ford was a young pitcher in 1950.

Comparing two pitchers is a little tougher than comparing two hitters since some of the standard pitching statistics are team-dependent. For example, if a pitcher has a great win/loss record such as 20–5, this could mean that he was an outstanding pitcher or it could mean that he was just a good pitcher whose team scored a lot of runs when he was pitching.

Here we consider the ERA pitching statistic, that is, the mean number of runs earned by the opposing team in a 9-inning game. An "earned" run is one that is not produced by means of an error by the team's fielders. This is one of the most common measures used to rate pitchers. Table 3.2 displays the season by season pitching statistics for both Roberts and Ford.

TABLE 3.2

Career pitching statistics for Robin Roberts and Whitey Ford.

Robin Roberts—born September 30, 1926

Year	Ag	Tm	Lg	W	L	G	IP	H	ER	HR	BB	SO	ERA	*lgERA	*ERA+
1948	21	PHI	NL	7	9	20	146.2	148	52	10	61	84	3.19	3.96	124
1949	22	PHI	NL	15	15	43	226.2	229	93	15	75	95	3.69	3.96	107
1950	23	PHI	NL	20	11	40	304.1	282	102	29	77	146	3.02	4.06	135
1951	24	PHI	NL	21	15	44	315.0	284	106	20	64	127	3.03	3.84	127
1952	25	PHI	NL	28	7	39	330.0	292	95	22	45	148	2.59	3.66	141
1953	26	PHI	NL	23	16	44	346.2	324	106	30	61	198	2.75	4.20	152
1954	27	PHI	NL	23	15	45	336.2	289	111	35	56	185	2.97	4.03	136
1955	28	PHI	NL	23	14	41	305.0	292	111	41	53	160	3.28	3.96	121
1956	29	PHI	NL	19	18	43	297.1	328	147	46	40	157	4.45	3.73	84
1957	30	PHI	NL	10	22	39	249.2	246	113	40	43	128	4.07	3.80	93
1958	31	PHI	NL	17	14	35	269.2	270	97	30	51	130	3.24	3.95	122
1959	32	PHI	NL	15	17	35	257.1	267	122	34	35	137	4.27	4.11	96
1960	33	PHI	NL	12	16	35	237.1	256	106	31	34	122	4.02	3.88	96
1961	34	PHI	NL	1	10	26	117.0	154	76	19	23	54	5.85	4.07	70
1962	35	BAL	AL	10	9	27	191.1	176	59	17	41	102	2.78	3.77	136
1963	36	BAL	AL	14	13	35	251.1	230	93	35	40	124	3.33	3.52	106
1964	37	BAL	AL	13	7	31	204.0	203	66	18	52	109	2.91	3.59	123
1965	38	BAL/HOU		10	9	30	190.2	171	59	18	30	97	2.78	3.42	123
1966	39	CHI/HOU		5	8	24	112.0	141	60	15	21	54	4.82	3.54	73

Whitey Ford—born October 21, 1928

Year	Ag	Tm	Lg	W	L	G	IP	H	ER	HR	BB	SO	ERA	*lgERA	*ERA+
1950	21	NYY	AL	9	1	20	112.0	87	35	7	52	59	2.81	4.30	153
1953	24	NYY	AL	18	6	32	207.0	187	69	13	110	110	3.00	3.68	123
1954	25	NYY	AL	16	8	34	210.7	170	66	10	101	125	2.82	3.42	121
1955	26	NYY	AL	18	7	39	253.7	188	74	20	113	137	2.63	3.76	143
1956	27	NYY	AL	19	6	31	225.7	187	62	13	84	141	2.47	3.87	156
1957	28	NYY	AL	11	5	24	129.3	114	37	10	53	84	2.57	3.60	140
1958	29	NYY	AL	14	7	30	219.3	174	49	14	62	145	2.01	3.54	176
1959	30	NYY	AL	16	10	35	204.0	194	69	13	89	114	3.04	3.63	119
1960	31	NYY	AL	12	9	33	192.7	168	66	15	65	85	3.08	3.60	117
1961	32	NYY	AL	25	4	39	283.0	242	101	23	92	209	3.21	3.70	115
1962	33	NYY	AL	17	8	38	257.7	243	83	22	69	160	2.90	3.73	129
1963	34	NYY	AL	24	7	38	269.3	240	82	26	56	189	2.74	3.52	128
1964	35	NYY	AL	17	6	39	244.7	212	58	10	57	172	2.13	3.62	170
1965	36	NYY	AL	16	13	37	244.3	241	88	22	50	162	3.24	3.39	105
1966	37	NYY	AL	2	5	22	73.0	79	20	8	24	43	2.47	3.33	135
1967	38	NYY	AL	2	4	7	44.0	40	8	2	9	21	1.64	3.13	191

```
         FORD   ROBERTS
                 1 |
              6 | 1 |
           4410 | 2 |
         988765 | 2 | 577799
          22000 | 3 | 001223
                 3 | 6
                 4 | 0024
                 4 | 8
                 5 |
                 5 | 8
```

FIGURE 3.5

Back-to-back stemplots of the season ERAs for Ford and Roberts.

Figure 3.5 displays back-to-back stemplots of the ERAs for Ford and Roberts. In constructing the stemplot, we break an ERA such as 2.81 at the decimal point, so the stem is 2 and the (one-digit) leaf is 8.

What do we see in comparing Ford's and Roberts' ERAs? Both pitchers had several seasons with ERAs in the 2.5–3.3 range. But Ford's worst season with respect to ERA was only 3.2 and he had five seasons where his ERA fell under 2.5. In contrast, Roberts's best season was 2.5 and he had six seasons where his ERA was over 4.0. Ford generally appears to be the better pitcher with respect to this pitching statistic.

But we have to be careful about making this conclusion. Why? Well, the measurement of a pitcher's ability, such as an ERA, should be made in the context of the season and league and ballpark in which the player played. Ford and Roberts played during roughly the same seasons. But Ford pitched primarily in the American League, Roberts in the National League, and they pitched in different ballparks. Maybe Ford's ERAs looked better than Roberts's ERAs because of the differences in league and ballpark.

To illustrate how much ERAs can be different across seasons and leagues, Figure 3.6 displays a time series plot of the ERA for all pitchers. The season ERAs of the National League are plotted using a light line and the AL ERAs by a darker line. Note the great rises and falls in the mean ERA across seasons. In the "dead-ball" era in the early 1900s, pitching was dominant and the mean ERA was low. Then pitching got worse over time and the AL ERA peaked at a value close to 5.0 around the year 1940. In the late 1960s, pitching again was dominant—the league ERAs dipped to about 3.0—and recently the ERAs have been rising. Also, note that the NL and AL ERAs were quite different for some seasons. Since the season ERAs are so variable, it is reasonable to view a pitcher's season ERA in the context of the average ERA that particular year.

Actually, the baseball-reference.com site goes one step further than a season adjustment. In the data tables shown above, the "lgERA" statistic is the average ERA of a pitcher in the same league and the same ballparks. We adjust a pitcher's ERA by computing

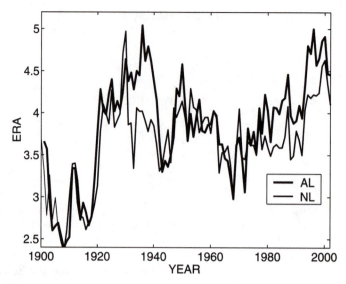

FIGURE 3.6
Time series plot of the season ERAs for all pitchers in the National and American Leagues.

$$^*\text{ERA} = \frac{\text{lgERA}}{\text{ERA}} \times 100$$

To illustrate the computation and interpretation of *ERA, Robin Roberts's ERA in 1950 was 3.02. The average ERA in the National League in the ballparks that Roberts pitched was 4.06. So Roberts's adjusted ERA was

$$^*\text{ERA} = \frac{4.06}{3.06} \times 100 = 135.$$

An adjusted ERA over 100 means that the pitcher performed above-average. Roberts had a good year in 1950—the league/ballpark average ERA was 35% higher than his ERA.

Suppose that we compare the adjusted ERAs of Roberts and Ford by stemplots in Figure 3.7.

We get some additional insight about the quality of these pitchers. Ford had an adjusted ERA of over 100 for every year that he pitched—that means that he was above average every year. In contrast, Roberts had 13 years where he was above-average and six years where he was below average. (It is interesting to note that five of his bad seasons occurred between 1956 and 1961.) Ford had three seasons where his adjusted ERA was 170 or higher. Ford does appear to be the superior pitcher in this study, but Ford pitched for a much stronger team (the Yankees) than Roberts (the Phillies), and it is possible that the difference in team strengths might explain part of the difference in ERAs.

```
          FORD   ROBERTS
                7 | 03
                8 | 4
                9 | 366
            5  10 | 67
          975  11 |
         9831  12 | 123347
            5  13 | 566
           30  14 | 1
           63  15 | 2
               16 |
           60  17 |
               18 |
            1  19 |
```

FIGURE 3.7
Back-to-back stemplots of the adjusted ERAs of Roberts and Ford.

Case Study 3-3: Home Runs—A Comparison of 1927, 1961, 1998, and 2001

Topics Covered: Stemplot, five-number summary, time series plot, boxplot   When a baseball fan thinks about home runs from a historical perspective, four years should come to mind: 1927, 1961, 1998, and 2001. The year 1927 was the year in which Babe Ruth hit 60 home runs and broke his own season record for home runs. This record stood for 34 years until 1961, when Roger Maris hit his 61 home runs. Another 37 years went by until Mark McGwire broke the season record with 70 home runs. It was generally thought that McGwire's record would last for a while, but Barry Bonds broke the record in 2001 (only three years later) with his great feat of 73 home runs.

Is Bonds really the greatest home run slugger of all time? Well, it is not clear. It is difficult to compare Bonds' accomplishment in 2001 with Ruth's accomplishment in 1927 since the difficulty of hitting a home run may have been different in the two years. Maybe we can better understand the magnitude of these players' accomplishments if we compare the home run hitting of all players for these four seasons.

Table 3.3 below gives the number of home runs (HR) hit by every team in the 1927, 1961, 1998, and 2001 baseball seasons. The table also gives the number of games (G) for each team each year. Note that the number of games in a season has changed over the years—the 1927 teams and the 1961 teams had season lengths that were about 7–9 games shorter than the current seasons. To make a reasonable comparison, it is helpful to compute the home run rate (RATE) that is the number of home runs hit divided by the number of games (G).

$$\text{RATE} = \frac{\text{HR}}{\text{G}}.$$

TABLE 3.3

Number of home runs and home run rates for all Major League teams in the years 1927, 1961, 1998, and 2001.

YEAR											
1927			**1961**			**1998**			**2001**		
HR	G	RATE	HR	G	RATE	HR	G	RATE	HR	G	RATE
54	156	0.35	158	154	1.03	147	162	0.91	208	162	1.28
84	153	0.55	157	154	1.02	214	162	1.32	174	162	1.07
109	155	0.70	183	155	1.18	205	162	1.27	194	162	1.20
74	153	0.48	188	155	1.21	198	163	1.21	176	162	1.09
29	153	0.19	103	155	0.66	198	162	1.22	213	162	1.31
39	154	0.25	128	154	0.83	165	162	1.02	166	162	1.02
37	155	0.24	176	156	1.13	134	163	0.82	208	162	1.28
57	155	0.37	103	155	0.66	115	163	0.71	206	162	1.27
158	155	1.02	240	163	1.47	207	162	1.28	209	162	1.29
56	155	0.36	180	163	1.10	149	162	0.92	131	162	0.81
29	157	0.18	149	163	0.91	234	161	1.45	147	162	0.91
51	156	0.33	138	163	0.85	111	163	0.68	164	162	1.01
36	153	0.24	150	161	0.93	201	162	1.24	161	162	0.99
26	153	0.17	112	163	0.69	221	163	1.36	161	162	0.99
55	155	0.35	167	161	1.04	159	162	0.98	235	162	1.45
28	154	0.18	189	162	1.17	215	162	1.33	199	162	1.23
			119	161	0.74	212	163	1.30	158	162	0.98
			90	162	0.56	138	162	0.85	136	161	0.84
						183	162	1.13	198	161	1.23
						114	162	0.70	214	162	1.32
						166	162	1.02	212	162	1.31
						159	162	0.98	139	162	0.86
						152	162	0.94	152	162	0.94
						147	162	0.91	164	162	1.01
						136	162	0.84	203	161	1.26
						126	163	0.77	199	162	1.23
						107	163	0.66	169	162	1.04
						167	162	1.03	121	162	0.75
						161	163	0.99	246	162	1.52
						223	163	1.37	195	162	1.20

This home run rate is also given in the table. If a team's RATE $= 1$, then it hits, on average, one home run each game.

One effective way of comparing the team home run rates across seasons is by means of parallel stemplots, shown in Figure 3.8. In the graph below, we divide the rate between the tenth and hundredth places, so a rate of 1.03 has a stem of 10 and a leaf of 3.

SEASON

1927		1961		1998		2001	
1	6888	1		1		1	
2	335	2		2		2	
3	24566	3		3		3	
4	8	4		4		4	
5	4	5	5	5		5	
6		6	668	6	58	6	
7	0	7	3	7	007	7	4
8		8	34	8	235	8	045
9		9	13	9	0013888	9	03799
10	1	10	123	10	123	10	112478
11		11	0268	11	2	11	9
12		12	1	12	12467	12	022267889
13		13		13	02256	13	012
14		14	7	14	5	14	5
15		15		15		15	1

FIGURE 3.8
Parallel stemplots of the team home run rates for the four seasons.

What do we see when we compare team home run rates across these four seasons?

- The pattern of home run hitting in 1927 is very interesting. Babe Ruth's team, the Yankees, averaged about a home run every game. But the Yankees were an outlier. A typical team in 1927 had a home run rate about .3—this average team hit a home run every three games. It is pretty clear that Ruth's 60 home run total was quite an accomplishment in the year 1927 where the rate of hitting home runs was small.

- The year 1961 was a sharp contrast to 1927 with respect to home run hitting. Most of the team home run rates are in the .8–1.1 range, which was comparable to the home run rate of the great 1927 Yankees. It was much more common to hit a home run in 1961 when Roger Maris hit his 61 roundtrippers.

- We see that the years 1998 and 2001 saw even greater home run hitting than 1961. Looking at the stemplots, it appears that a typical home run rate has increased from 1961 to 1998 and from 1998 to 2001. Half of the 2001 baseball teams had home run rates of 1.19 or higher.

TABLE 3.4
Five-number summaries of the team home rates for the four seasons.

Year	N	LO	Q1	Median	Q3	HI
1927	16	0.16	0.20	0.34	0.45	1.01
1961	18	0.55	0.73	0.98	1.14	1.47
1998	30	0.65	0.85	1.00	1.27	1.45
2001	30	0.74	0.99	1.14	1.28	1.51

To get a better handle on the differences in home run hitting between seasons, we can compute summary statistics. Table 3.4 gives five-number summaries for each batch of team home run rates. Figure 3.9 draws parallel boxplots using these summaries.

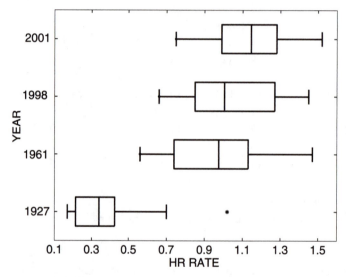

FIGURE 3.9
Parallel boxplots of the team home run rates for the four seasons.

This display reinforces the conclusions that were drawn earlier. Home runs were generally not hit in 1927 and there was a small spread in the team home run rates. The dot in the 1927 boxplot indicates the outlying team, the Yankees. From 1961 to 2001 there has been a gradual increase in the rate of hitting home runs. Comparing the median home run rate (.98) in 1961 with the median rate (1.14) in 2001, we see that a typical team in 2001 hit approximately $1.14 - .98 = .16$ more home runs than a typical team in 1961—that translates to about an additional home run each six games.

Case Study 3-4: Slugging Percentages Are Normal

Topics Covered: Stemplot, mean and standard deviation, normal distribution probabilities
The standard deviation, denoted s, is a measure of dispersion about the center (mean) and has a useful interpretation when our data is bell-shaped. We illustrate the use of the empirical rule or "68–95–99.7 rule" that tells us about the fraction of observations that fall within three intervals.

Let's look at all hitters in 1999 that had at least 400 at-bats—we can regard these hitters as the "regular" players, since they played most of the games during the season. I drew a MINITAB stemplot, shown in Figure 3.10, of the slugging percentages (SLG) for the 178 regular players.

```
Stem-and-leaf of SLG    N = 178
     Leaf Unit = 0.010
                    2 | 7
                    2 | 9
                    3 | 001
                    3 | 2
                    3 | 4555
                    3 | 666667777
                    3 | 88999999
                    4 | 0000000111111111
                    4 | 222222222233333333333
                    4 | 444444455555555555555
                    4 | 66666666667777777777
                    4 | 8888888899999
                    5 | 00000011111
                    5 | 22222222233333333
                    5 | 444444455555555555
                    5 | 6667777
                    5 | 8889
                    6 | 0001
                    6 | 333
                    6 |
                    6 | 6
                    6 | 9
                    7 | 1
```

FIGURE 3.10
Stemplot of slugging percentages for the 178 regular players in the 1999 season.

This set of slugging percentages looks pretty symmetric; most of the values are in the .400 to low .500 range and there is a large spread—the best hitter in 1999 had a slugging percentage over .700 and the worst was only about .270. (Who were the players with the best and worst SLG values in 1999? It might be easier to guess the largest value rather than the smallest.)

We use MINITAB to compute the mean and standard deviation for these SLGs.

Descriptive Statistics

Variable	N	Mean	Median	TrMean	StDev	SEMean
SLG	178	0.47313	0.46550	0.47224	0.07757	0.00581

Variable	Min	Max	Q1	Q3
SLG	0.27600	0.71000	0.42075	0.52975

Rounding, we see that $\bar{x} = .473$ and $s = .078$.

When the distribution of the data is bell-shaped (as it is here), we expect

- 68% of the data to fall within one standard deviation of the mean (that is, between $\bar{x} - s$ and $\bar{x} + s$),

- 95% of the data to fall within 2 standard deviations of the mean (between $\bar{x} - 2s$, $\bar{x} + 2s$),

- 99.7% of the data to fall within 3 standard deviations of the mean (between $\bar{x} - 3s$, $\bar{x} + 3s$).

Applying this 68–95–99.7 rule, we compute

$$\bar{x} - s \quad \text{and} \quad \bar{x} + s,$$
$$.473 - .078 \quad \text{and} \quad .473 + .078,$$
$$.395 \quad \text{and} \quad .551.$$

We expect 68% of the SLG's to fall between .395 and .551.

Let's check how many actually fall between .395 and .551: there are 126 SLGs between .395 and .551 (using the original dataset) and the proportion of SLGs in this interval is $126/178 = .708$, which is pretty close to .68.

Next, we compute

$$\bar{x} - 2s \quad \text{and} \quad \bar{x} + 2s,$$
$$.473 - 2(.078) \quad \text{and} \quad .473 + 2(.078)$$
$$.317 \quad \text{and} \quad .629.$$

We expect 95% of the SLG's to fall between .317 and .629. Checking the data again, we obtain 168 (out of 178) in this interval, which corresponds to $168/178 = .944$, which is very close to .95.

Last, we compute

$$\bar{x} - 3s \quad \text{and} \quad \bar{x} + 3s,$$
$$.473 - 3(.078) \quad \text{and} \quad .473 + 3(.078),$$
$$.239 \quad \text{and} \quad .707.$$

We expect practically all (99.7%) of the SLG's to fall between .239 and .707. In fact 177 (out of 178) fall in this range, which is $177/178 = .994$, which is very close to .997. So again this rule seems to work.

Generally, most "derived" baseball statistics that are computed by a division or a multiplication will be bell-shaped and well-suited for applying the 68–95–99.7 rule. Examples of such derived statistics are AVG, SLG, OBP, and ERA.

Case Study 3-5: Great Batting Averages

Topics Covered: Stemplot, mean and standard deviation, standardized score In this case study, we examine great batting averages. We identify four players who had unusually high batting averages (AVG) in recent history:

- 1994: Tony Gwynn (SD) hit .394

- 1980: George Brett (KC) hit .390

- 1977: Rod Carew (MIN) hit .388

- 1941: Ted Williams (BOS) hit .406

Now, on face value, any baseball statistics like these high averages are meaningless because we don't know the context in which these statistics were achieved. If I tell you that Joe Schmoe (a hypothetical player) batted .420, you should ask:

- When? You have to know when Joe Schmoe got this AVG. For example, batting averages in 1900 were generally much higher than batting averages in 1968.

- Where? It matters where Joe played. Some parks (like Coors Field) are easier to hit in, and others (like Dodger Stadium) are harder to hit in.

- How many at-bats? We'll talk about this in a later chapter, but it is easier to hit .420 based on 100 AB than 500 AB.

We can make better sense of a batting average, like Carew's .388 AVG in 1977, when we see it compared to the AVGs of all players in the year 1977.

Let's focus on the 1977 hitters who had at least 400 at-bats. Figure 3.11 displays a stemplot of the 168 AVGs.

Here are some features of this dataset. (1) The distribution looks symmetric and bell-shaped. (2) A typical AVG is about .275. (3) Most of the averages fall between .230 and .300. (4) We can't miss the one large AVG at .388, which corresponds to Carew.

We would like a measure of *relative standing* that tells us how good Carew's batting average is. We compute a *standardized score* of an average by subtracting the mean $\bar{x}$ and then dividing by the standard deviation s.

Calculations reveal that $\bar{x} = .27742$ and $s = .02717$ for these 1977 AVGs. So the standardized score (or z score) for Carew's .388 is

$$z = \frac{.388 - .27742}{.02717} = 4.07 \qquad \text{(WOW!)}$$

Stem-and-leaf of 1977 $N = 168$
Leaf Unit $= 1.0$

```
20 | 06
21 | 6
22 | 9
23 | 0133559
24 | 001135566778
25 | 00111222344567799
26 | 0011112233344445556677899
27 | 00112223344555557788999
28 | 000001222233334444446777788999
29 | 0001111122336778899
30 | 00022445778899
31 | 011225788
32 | 002568
33 | 668
34 |
35 |
36 |
37 |
38 | 8 (CAREW)
```

FIGURE 3.11
Stemplot of batting averages of the 1977 Major League players with at least 400 at-bats.

Carew's z-score is 4.07, which means that his AVG was about four standard deviations *above* the mean.

Let's compute the standardized score for the weakest hitter in 1977 who only batted .200. His z-score would be

$$z = \frac{.200 - .27742}{.02717} = -2.85.$$

In other words, his AVG was 2.85 standard deviations *below* the mean.

If a player has a z-score of 0, this means that his AVG is equal to the mean .277. So

- A positive z-score corresponds to an AVG above the mean $\bar{x}$.

- A negative z-score corresponds to an AVG below the mean $\bar{x}$.

- A z-score of 0 corresponds to an AVG that is equal to the mean $\bar{x}$.

Recall, for bell-shaped data, practically all of the z-scores for the data will fall between -3 and 3. (This is a restatement of the 99.7 rule.)

Using the concept of standardized scores, we can compare these great batting averages:

- We already saw that Carew's .388 corresponded to a z-score of 4.07.

- Tony Gwynn's .394—that year (1994), the mean $\bar{x} = .29311$ and $s = .03201$, so his z-score would be

$$z = \frac{.394 - .29311}{.03201} = 3.15.$$

- Ted Williams's .406—that year (1941), $\bar{x} = .28063$ and $s = .03281$, and

$$z = \frac{.406 - .28063}{.03281} = 3.82.$$

- George Brett's .390—that year (1980), $\bar{x} = .27882$ and $s = .02757$, and

$$z = \frac{.390 - .27882}{.02757} = 4.03.$$

Comparing the four, Carew's AVG was best (relative to his peers) since he had the largest standardized score.

In this comparison, we see that batting averages have changed over the years. Table 3.5 shows the mean and standard deviation for five years:

TABLE 3.5
Mean and standard deviation of players' batting averages with at least 400 at-bats for five years.

Year	Mean $\bar{x}$	Standard Deviation s
1900	.295	.037
1941	.280	.033
1977	.277	.027
1980	.279	.028
1994	.293	.032

A couple of general comments from this table:

- Looking at the means, there doesn't seem to be any general trend in batting averages— they have gone up and down. Higher batting averages could mean better hitting or weaker pitching, or both.

- If you look at the standard deviations over years, you will see a general decreasing pattern. The standard deviation of AVGs today is smaller than it used to be. This means that the hitting abilities of the current players are more similar than they used to be.

Exercises

Leadoff Exercise 1. Rickey Henderson and Tim Raines were both great leadoff hitters who played in the 1980's and 1990's. Table 3.6 shows career on-base percentage (OBP)

TABLE 3.6

On-base and slugging percentages for Rickey Henderson and Tim Raines for the seasons of their career.

	Henderson			Raines	
Age	OBP	SLG	Age	OBP	SLG
20	.338	.336			
21	.420	.399	21	.391	.438
22	.408	.437	22	.353	.369
23	.398	.382	23	.393	.429
24	.414	.421	24	.393	.437
25	.399	.458	25	.405	.475
26	.419	.516	26	.413	.476
27	.358	.469	27	.429	.526
28	.423	.497	28	.350	.431
29	.394	.399	29	.395	.418
30	.411	.399	30	.379	.392
31	.439	.577	31	.359	.345
32	.400	.423	32	.380	.405
33	.426	.457	33	.401	.480
34	.432	.474	34	.365	.409
35	.411	.365	35	.374	.422
36	.407	.447	36	.383	.468
37	.410	.344	37	.403	.454
38	.400	.342	38	.395	.383
39	.376	.347			
40	.423	.466			
41	.368	.305			
42	.366	.351			

and slugging percentage (SLG) for each player—the seasons where Raines had less than 200 at-bats have been removed.

(a) Construct back-to-back stemplots of the OBPs for Henderson and Raines.

(b) Find five-number summaries of the two batches.

(c) Based on your work in (a) and (b), which hitter was more effective in getting on-base? On average, how much better was the superior player?

(d) Construct one time series plot of the slugging percentages of Henderson and Raines graphing against age. Compare the patterns of change (across time) for each player. Was one player consistently better than the other with respect to SLG? Did both players display the typical pattern in which one gets better in performance, hits a peak, and then declines towards the end of his career?

Leadoff Exercise 2. If you look at Rickey Henderson's batting statistics in Table 1.1, we see that his highest season OBP was .439 in the 1990 season. How impressive was that .439 OBP in the 1990 season? To check, Figure 3.12 displays a stemplot of the OBPs of all players that had at least 400 plate appearances in the 1990 season. A box is placed around Rickey's value.

```
25 | 88
26 | 7
27 | 466999
28 | 2222589
29 | 0012236799
30 | 00112334444445557778
31 | 223466889
32 | 000122234466677888999
33 | 000002233345666778899
34 | 000111222223467788899
35 | 001223466777899
36 | 00444466679
37 | 012345555566789
38 | 13455679
39 | 77
40 | 0457
41 | 377
42 |
43 | 9
44 | 1
```

FIGURE 3.12
Stemplot of the OBPs of all players with at least 400 plate appearances in 1990.

(a) Describe the general shape of the distribution of OBPs and discuss any unusual features.

(b) The mean and standard deviation of the OBPs are given by $\bar{x} = .337$ and $s = .036$, respectively. Find the standardized score of Rickey's OBP of .439.

(c) Find an interval that contains approximately 95% of the OBPs.

(d) How does Rickey's .439 OBP compare with the "greatness" of the batting averages discussed in Case Study 3-5. Is this study sufficient to convince you that Rickey was the greatest leadoff hitter of all time? Explain.

3.1. (Comparing batting averages across eras)   The stemplots in Figure 3.13 display the batting averages of all players with at least 400 at-bats in the years 1897 and 1997.

1897 players		1997 players
0	20	2
	21	
	22	389
	23	35799
	24	012334567889
	25	000000222234678
76	26	01112333445555557789999
73	27	0112333345555677789
9997554322	28	000112234456666667788
210	29	000001111234444455667788899
876520	30	0223445678888
8865320	31	01344889
9998521	32	1337789
75	33	022
53000	34	7
54322	35	
21	36	16
7	37	1
92	38	
	39	
	40	
	41	
3	42	

FIGURE 3.13
Stemplots of batting averages of all players with at least 400 at-bats from 1897 and 1997 seasons.

(a) Find the five-number summaries of each dataset.

(b) Which group of players tends to have the higher batting averages? Explain.

(c) Which group of players had the greater spread in batting averages? Explain.

3.2. (Continuation of Exercise 3.1)

(a) For the 1897 batting averages, one can compute the mean and standard deviation to be .318 and .037, respectively. Find an interval of values where you expect 68% of the batting averages to fall.

(b) Count the number of players who had a batting average in the interval in (a).

(c) Find the proportion of players who had an average in the interval in (a). Is it close to what you expect?

(d) For the 1997 batting averages, the mean = .281 and the standard deviation = .028. Find an interval where you think 95% of the averages will fall.

(e) Find the proportion of 1997 players who had an average in the interval you found in (d). Check to see if it is close to what you expect.

3.3. (Continuation of Exercise 3.1)

(a) In 1897, Willie Keeler won the batting crown with an average of .424. Using the mean and standard deviation of batting averages given in the previous exercise, find the standardized score for Keeler's average.

(b) In 1997, Tony Gwynn won the batting crown with an average of .372. Find the standardized score for Gwynn's average.

(c) Based on your computations in (a) and (b), which batter was better relative to his contemporaries? Why?

(d) In the book *Full House*, Stephen Jay Gould argues that ballplayers are generally getting less variable over time and it will be difficult for a player to hit for a .400 season batting average in the future. On the basis of your computations, do you agree with Gould's assertion? Why?

3.4. (Comparison of hit profiles)   Table 3.7 gives the number of at-bats, hits, doubles, triples and home runs for the players during the 1900, 1945, and 1990 baseball seasons.

TABLE 3.7
Batting statistics for all players in the 1900, 1945, and 1990 baseball seasons.

YEAR	AB	H	1B	2B	3B	HR
1900	39132	10925		1432	607	254
1945	84447	21977		3497	728	1007
1990	142768	36817		6526	865	3317

(a) For each year, compute the number of singles and put the numbers in the 1B column.

(b) For each year, compute the proportion of hits that are doubles. (For 1900, for example, you would divide the number of doubles 1432 by the number of hits 10925.) Put your answers in the table below.

YEAR	Proportion of hits that are			
	1B	2B	3B	HR
1900				
1945				
1990				

(c) For each year, compute the proportion of hits that are singles, proportion of hits that are triples, and the proportion of hits that are home runs.

(d) Based on the proportions you computed in (b) and (c), comment on how the game of baseball has changed from 1900 to 1990. Would you say that the game is more or less exciting than the games in the past? Why?

3.5. (Comparisons of rates of strikeouts and walks) Table 3.8 gives the number of at-bats, strikeouts, and walks for the three baseball seasons 1900, 1945, and 1990.

TABLE 3.8
Strikeout and walk statistics for all players in the 1900, 1945, and 1990 baseball seasons.

Year	AB	BB	SO
1900	39132	3034	2697
1945	84447	8295	8050
1990	142768	13852	24390

(a) For each season, compute the number of plate appearances by adding the at-bats to the walks. Put your answers in the PA column.

YEAR	PA	Proportion of	
		BB	SO
1900			
1945			
1990			

(b) For each year, compute the proportion of PAs that are strikeouts.

(c) For each year, compute the proportion of PAs that are walks.

(d) Explain how pitching has changed from 1900 to 1990 by looking at the proportions you computed in (b) and (c). Have pitchers become more or less dominating over the years? Do pitchers today have more or less control?

3.6. (Griffey vs. Bonds, continued) Refer back to Case Study 3-1 comparing the hitting of Junior Griffey and Barry Bonds.

(a) Construct back-to-back stemplots of the season on-base percentages (OBP) for Bonds and Griffey.

(b) Compare the two datasets. Which hitter generally has a higher OBP? Which hitter has the larger variation in OBP from season to season?

(c) Repeat (a) and (b) using the season slugging percentages (SLG).

(d) Based on your work, who is the more valuable hitter—Bonds or Griffey? Explain why.

(e) Although Bonds may have better hitting statistics than Griffey, some people still claim that Griffey is the more valuable player. Why might they say that?

3.7. Table 3.9 gives the number of wins (W), losses (L), winning proportion (PCT), and Earned Run Average (ERA) for two great modern pitchers, Greg Maddux and Tom Glavine.

(a) Using back-to-back stemplots, compare the season winning percentages of Maddux and Glavine.

TABLE 3.9
Career pitching statistics for Greg Maddux and Tom Glavine.

	Greg Maddux				Tom Glavine			
Year	W	L	PCT	ERA	W	L	PCT	ERA
1986	2	4	0.333	5.52				
1987	6	14	0.300	5.61	2	4	0.333	5.54
1988	18	8	0.692	3.18	7	17	0.291	4.56
1989	19	12	0.612	2.95	14	8	0.636	3.68
1990	15	15	0.500	3.46	10	12	0.454	4.28
1991	15	11	0.576	3.35	20	11	0.645	2.55
1992	20	11	0.645	2.18	20	8	0.714	2.76
1993	20	10	0.666	2.36	22	6	0.785	3.2
1994	16	6	0.727	1.56	13	9	0.590	3.97
1995	19	2	0.904	1.63	16	7	0.695	3.08
1996	15	11	0.576	2.72	15	10	0.600	2.98
1997	19	4	0.826	2.2	14	7	0.666	2.96
1998	18	9	0.666	2.22	20	6	0.769	2.47
1999	19	9	0.678	3.57	14	11	0.560	4.12
2000	19	9	0.678	3.00	21	9	0.700	3.40
2001	17	11	0.607	3.05	16	7	0.696	3.57

(b) Based on your comparison, which pitcher tended to win a greater percentage of games?

(c) Compare the season ERAs of Maddux and Glavine using stemplots.

(d) Which pitcher tended to have a lower season ERA?

3.8. Table 3.10 shows the slugging percentage (SLG) for the "regular" shortstops and "regular" 3rd basemen in 1998.

(a) Construct parallel boxplots of the slugging percentages of the shortstops and the 3rd basemen.

(b) Based on your work in (a), which type of player is more likely to be a slugger?

(c) On the average, how much superior is one group over the other group with respect to slugging percentage ?

(d) Is the general pattern you found in (b) and (c) true for all players? Find some players that don't agree with the general pattern.

TABLE 3.10
Slugging percentages for regular shortstops and 3rd basemen in 1998.

Shortstops		3rd Basemen	
Last Name	SLG	Last Name	SLG
AURILIA	.406	WILLIAMS	.439
RENTERIA	.342	BOGGS	.400
CARUSO	.390	BLOWERS	.386
CRUZ	.354	CAMINITI	.508
GUTIERREZ	.334	COOMER	.406
RELAFORD	.338	RANDA	.367
GARCIAPARRA	.584	DAVIS	.442
LARKIN	.503	ALFONZO	.427
GOMEZ	.378	MUELLER	.395
MEARES	.368	FRYMAN	.504
JETER	.480	HERNANDEZ	.471
VIZQUEL	.371	BROSIUS	.471
BORDICK	.410	PALMER	.510
VALENTIN	.392	VALENTIN	.442
ORDONEZ	.299	CIRILLO	.445
BELL	.431	JONES	.547
DiSARCINA	.384	ROLEN	.532
GONZALEZ	.360	VENTURA	.435
RODRIGUEZ	.559	RIPKEN	.389
PEREZ	.381	CASTILLA	.589

3.9. (Triple rates for 1899 and 1999 players)   Old-timers would argue that there is less use of speed in baseball today than in the old days. One indication of the lack of speed in baseball today is the relatively small number of triples hit. (Other explanations for the small number of triples include the liveliness of the ball and the change in ball park design.) For each of the "regular" players (with at least 400 at-bats) in 1899 and 1999, we compute the triple rate = 3B/AB. Stemplots of the triple rates for each group of players are shown in Figure 3.14. For the stemplots, the break point is between the hundredths and thousandths places, so for a triple rate of .048, the stem is 4 and the leaf is 8.

(a) Compute five-number summaries of each dataset and graph parallel boxplots.

(b) Which group of players tended to get a higher rate of triples? What is the difference between the two average triple rates?

(c) Does your comparison support the claim that there is less speed in baseball nowadays? What other variables could you use to measure speed of ballplayers?

1899 players

```
0 | 114
0 | 6667999
1 | 00011222233334444
1 | 55555666788889999
2 | 01122233334
2 | 68
3 | 0012
3 | 7
4 | 23
```

1999 players

```
0 | 000000000000000000000000000001111111111111111111
0 | 2222222222223333333333333333
0 | 44444444444444455555555
0 | 66666666666666667777777
0 | 8888888888888889999999
1 | 0000000000111
1 | 222222333
1 | 44444455
1 | 666
1 | 889
2 | 0
```

FIGURE 3.14
Stemplots of triple rates of players in 1899 and 1999 seasons with at least 400 at-bats.

3.10. Table 3.11 displays the total number of home runs (HR) hit by each of the 30 major league teams in 1999.

(a) Using the stems shown below, construct back-to-back stemplots of the HR totals of the American League and the National League.

TABLE 3.11
Home run numbers of all Major League teams in 1999.

American League		National League	
Team	HR	Team	HR
MIN	105	FLA	128
KC	151	CHI	189
TB	145	MON	163
DET	212	COL	223
ANA	158	SD	153
CHW	162	MIL	165
BAL	203	STL	194
SEA	244	LA	187
TOR	212	PHI	161
OAK	235	PIT	171
BOS	176	SF	188
TEX	230	CIN	209
CLE	209	NY	181
NYY	193	HOU	168
		AZ	216
		ATL	197

American League	National League
	10
	11
	12
	13
	14
	15
	16
	17
	18
	19
	20
	21
	22

(b) Find five-number summaries of the two datasets.

(c) Using your work in (a) and (b), compare the two datasets. Did one league tend to hit more home runs than the other league? Which teams hit an unusually small or large number of home runs in 1999?

3.11. (Comparing two hitters)   Find two contemporary hitters who have each played ten or more major league seasons. Compare the two hitters on the basis of season data using one batting statistic such as AVG, SLG, HR, or OBP. You should (a) list the season data for both players, (b) compare the datasets using appropriate graphs, (c) compute summaries of both datasets, and (d) describe what you've learned from this comparison.

3.12. (Comparing two pitchers)   Find two contemporary pitchers who have each played ten or more major league seasons. Compare the two pitchers on the basis of season data using one pitching statistic such as ERA, PCT, SO, or BB. You should (a) list the season data for both players, (b) compare the datasets using an appropriate graph, (c) compute summaries of both datasets, and (d) describe what you've learned from this comparison.

3.13. (Comparing two teams)   Find two teams that you are interested in comparing with respect to some batting statistic, such as AVG, SLG, OBP, or HR. For each of the two teams, list all of the regular players on the team for a particular season (the ones with at least 400 at-bats) and the value of the batting statistic for each player. Compare the two batches of data using an appropriate graph, compute summaries of both batches, and describe which team appears to be superior from the viewpoint of this batting statistic.

3.14. (Comparing NL and AL teams)   Compare the NL and AL teams for any season with respect to a particular hitting, pitching, or fielding statistic. List the 30 teams and the value of the statistic for each team. Compare the NL and AL batches using a graph, compute statistics, and describe which league appears to be superior with respect to this statistic.

3.15. (Comparing players from different eras)   Compare baseball hitters or pitchers from two different years. Choose two years of interest (not too close together) and one baseball statistic that you are interested in. Randomly pick at least 30 regular players from each year, and list the name and the baseball statistic for each player. Compare the two years of data using an appropriate graph, compute summary statistics, and describe what you have learned in this comparison.

3.16. There has been a lot of talk about the recent surge in home run hitting. How has the rate of home run hitting changed since 1990? From a baseball web site, find the total number of home runs and at-bats for each of the 30 teams for the current year and compute the home run rate for the teams. Find the home run rates for all the 1990 teams. (This data is available on the book website described in Appendix 2.) Comparing the 1990 and current datasets, has there been a significant change in the rate of home run hitting?

Further Reading

Devore and Peck (2000) and Moore and McCabe (1998) provide good descriptions of the exploratory methods used in this chapter to compare several batches of data. Chapter 2 of Albert and Bennett (2001) illustrates the use of these methods to compare batches of baseball data.

4

Relationships Between
Measurement Variables

What's On-Deck?

This chapter illustrates statistical methods for studying the association between different baseball statistics. A large number of statistics, such as hits, doubles, triples, home runs, batting average, slugging percentage, and on-base percentage are used to evaluate hitters. Case Study 4-1 explores the relationships between these different batting statistics for the teams in the 2000 baseball season. The basic tool used in this case study is the scatterplot, and the patterns in the scatterplots are informative in assessing the strength and direction of association between pairs of variables. The most valuable team batting statistic is the one that is most highly associated with runs scored per game. In Case Study 4-2, we rank the different batting statistics with respect to their correlation with runs scored. In Case Study 4-3, we go one step further and evaluate a number of batting statistics, such as batting average, slugging percentage, on-base percentage, and OPS (on-base percentage plus slugging percentage) by how close each statistic can be used to predict teams' runs scored per game. Multiple regression is a useful tool for finding the "best" batting statistic for predicting team runs based on the number of singles, doubles, triples, home runs, walks, and hit-by-pitch by the team. In Case Study 4-4, we find the best linear combination of these batting events using the least-squares criterion, and the weights of this linear combination are useful for comparing the worth of a single, double, triple, and home run from the standpoint of producing runs.

The remaining case studies illustrate some interesting relationships in baseball data. In Case Study 4-5 we illustrate Bill James's Pythagorean Formula, which is an empirical relationship between the ratio of a team's wins and losses and the square of the ratio of a team's runs scored and runs allowed. In Case Study 4-6, we see that there is a natural tendency for a player's statistic one year to regress toward the average value of the statistic for the next year. Despite popular opinion, it is natural for a hot-hitting rookie to have a less-than-hot hitting performance the following year. Last, in Case Study 4-7, we explore the "2000 Dinger Drop-off" where there was a tendency for baseball teams to hit fewer home runs in the second half of the 2000 baseball season.

Case Study 4-1: Relationships in Team Offensive Statistics

Topics Covered: Relationships between two measurement variables, scatterplot, looking for association   In this case study we discuss relationships in baseball hitting data. Table 4.1 lists batting statistics for all 30 teams for the 2000 baseball season. These data

TABLE 4.1
Team names, league, and batting statistics for the teams in the 2000 baseball season.

Team	LG	AVG	OBP	SLG	AB	R	H	2B	3B	HR	RBI	BB	SO	HBP
Anaheim	Am	0.280	0.352	0.472	5628	864	1574	309	34	236	837	608	1024	47
Baltimore	Am	0.272	0.341	0.435	5549	794	1508	310	22	184	750	558	900	49
Boston	Am	0.267	0.341	0.423	5630	792	1503	316	32	167	755	611	1019	42
Chicago	Am	0.286	0.356	0.470	5646	978	1615	325	33	216	926	591	960	53
Cleveland	Am	0.288	0.367	0.470	5683	950	1639	310	30	221	889	685	1057	51
Detroit	Am	0.275	0.343	0.438	5644	823	1553	307	41	177	785	562	982	43
Kansas City	Am	0.288	0.348	0.425	5709	879	1644	281	27	150	831	511	840	48
Minnesota	Am	0.270	0.337	0.407	5615	748	1516	325	49	116	711	556	1021	35
New York	Am	0.277	0.354	0.450	5556	871	1541	294	25	205	833	631	1007	57
Oakland	Am	0.270	0.360	0.458	5560	947	1501	281	23	239	908	750	1159	52
Seattle	Am	0.269	0.361	0.442	5497	907	1481	300	26	198	869	775	1073	48
Tampa Bay	Am	0.257	0.329	0.399	5505	733	1414	253	22	162	692	559	1022	50
Texas	Am	0.283	0.352	0.446	5648	848	1601	330	35	173	806	580	922	39
Toronto	Am	0.275	0.341	0.469	5677	861	1562	328	21	244	826	526	1026	60
Arizona	Na	0.265	0.333	0.429	5527	792	1466	282	44	179	756	535	975	59
Atlanta	Na	0.271	0.346	0.429	5489	810	1490	274	26	179	758	595	1010	59
Chicago	Na	0.256	0.335	0.411	5577	764	1426	272	23	183	722	632	1120	54
Cincinnati	Na	0.274	0.343	0.447	5635	825	1545	302	36	200	794	559	995	64
Colorado	Na	0.294	0.362	0.455	5660	968	1664	320	53	161	905	601	907	42
Florida	Na	0.262	0.331	0.409	5509	731	1441	274	29	160	691	540	1184	60
Houston	Na	0.278	0.361	0.477	5570	938	1547	289	36	249	900	673	1129	83
Los Angeles	Na	0.257	0.341	0.431	5481	798	1408	265	28	211	756	668	1083	51
Milwaukee	Na	0.246	0.325	0.403	5563	740	1366	297	25	177	708	620	1245	61
Montreal	Na	0.266	0.326	0.432	5535	738	1475	310	35	178	705	476	1048	29
New York	Na	0.263	0.346	0.430	5486	807	1445	282	20	198	761	675	1037	45
Philadelphia	Na	0.251	0.329	0.400	5511	708	1386	304	40	144	668	611	1117	44
Pittsburgh	Na	0.267	0.339	0.424	5643	793	1506	320	31	168	749	564	1032	66
San Diego	Na	0.254	0.330	0.402	5560	752	1413	279	37	157	714	602	1177	46
San Francisco	Na	0.278	0.362	0.472	5519	925	1535	304	44	226	889	709	1032	51
St. Louis	Na	0.270	0.356	0.455	5478	887	1481	259	25	235	841	675	1253	84

give GP, AB, R, H, 2B, 3B, HR, RBI, AVG, TB, SLG, and OBP for all teams. When there are many variables measured on each team, we are interested in exploring relationships among them.

Relating Slugging Percentage and On-Base Percentage

A first step in examining the relationship between two numerical variables, say SLG and OBP, is to draw a scatterplot. This graph plots each pair of values on a grid, and we look for association by finding particular patterns in the display.

Suppose one team has an SLG value of .430 and an OBP value of .333. Here we've decided to use OBP on the vertical scale and SLG on the horizontal. (It would not have mattered if we had switched what variables were on which axis.) We plot (SLG = .430, OBP = .333) by plotting the point (.430, .333) on a Cartesian grid. We continue for the remaining 29 teams, getting the display shown in Figure 4.1.

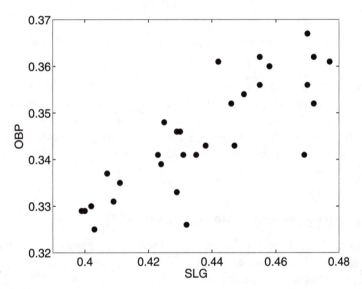

FIGURE 4.1
Scatterplot of slugging percentage and on-base percentage for the 2000 Major League teams.

We see a general tendency for the points to drift from (left, low) to (right, high), indicating that teams with a low OBP tend also to have a low SLG, and teams with high OBP tend to have a high SLG. This is a positive association in the scatterplot.

Relating Triples and Doubles

As a second example, we draw a labeled scatterplot, shown in Figure 4.2, of the number of triples and number of doubles hit by the 30 teams. Each plotting point has an abbreviation label for the team.

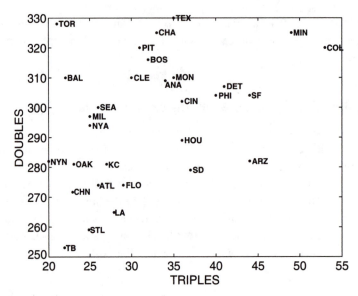

FIGURE 4.2
Scatterplot of numbers of doubles and triples for the 2000 Major League teams.

The team labels give some interesting information: Colorado (COL) was high in both triples and doubles and Tampa Bay (TB) was low on both variables. Looking at the pattern in the graph, we don't see a very strong association between the numbers of doubles and triples hit. The points may be drifting slightly upward as one goes from left to right, but the positive relationship is clearly weaker than the relationship we saw above between OBP and SLG.

Relating Home Runs and Triples for Historical Teams

To illustrate a different type of relationship, 30 teams were randomly selected from the group of all major league baseball teams in the 100-year period from 1901 to 2000. For each team, we collected two hitting statistics:

- Home run rate (HR/G): the number of home runs hit per game. This statistic is computed by dividing the number of team home runs by the number of games played.

- Triple rate (3B/G): the number of triples hit per game. This is computed by dividing the number of triples by the number of games.

Table 4.2 gives the team names, the years, and the two hitting statistics.

In the scatterplot in Figure 4.3, we plot the home run rate against the triple rate for the 30 teams.

Here the points drift from top left to bottom right, which corresponds to a negative relationship in the graph. Teams that hit a high rate of home runs tended to hit a low rate of triples, and teams that hit a low rate of home runs tended to have a high triple rate. We can explain this relationship if we look at the years of the teams in the table. In Figure 4.4,

TABLE 4.2
Team names, years, home runs per game, and triples per game for 30 historical baseball teams.

Team	Year	HR/G	3B/G	Team	Year	HR/G	3B/G
DET	1901	0.21	0.59	OAK	1971	0.99	0.16
NY1	1906	0.10	0.35	DET	1974	0.81	0.22
DET	1907	0.07	0.49	CAL	1975	0.34	0.25
WS1	1908	0.05	0.48	HOU	1980	0.46	0.41
CHI	1911	0.34	0.64	ATL	1983	0.80	0.28
PHI	1913	0.46	0.49	CHI	1984	0.84	0.29
KCF	1914	0.25	0.50	MON	1985	0.73	0.30
DET	1918	0.10	0.44	CIN	1986	0.89	0.22
SLA	1935	0.47	0.33	TEX	1986	1.14	0.27
WS1	1940	0.34	0.44	CIN	1988	0.76	0.16
CHI	1940	0.56	0.31	SEA	1993	0.99	0.15
CLE	1944	0.45	0.32	PIT	1996	0.85	0.20
PHI	1955	0.86	0.32	MIL	1997	0.84	0.17
BOS	1957	0.99	0.21	BAL	1998	1.32	0.07
DET	1964	0.96	0.35	BAL	2000	1.14	0.14

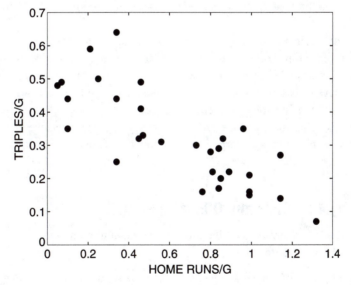

FIGURE 4.3
Scatterplot of home run rates against triple rates for 30 historical baseball teams.

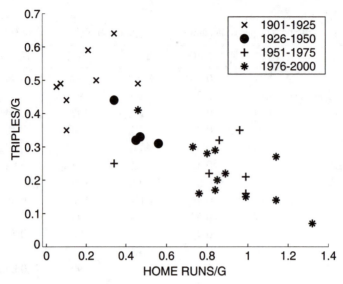

FIGURE 4.4
Scatterplot of home run rates against triple rates for 30 historical baseball teams where the plotting symbol gives the era of the team.

we have redrawn the scatterplot, where the plotting symbol corresponds to the time when the team played.

Note that the points in the upper left portion of the scatterplot generally correspond to teams that played in the first quarter of the 20th century, and the points in the lower right section correspond to teams that played in the most recent period from 1976–2000. In the early days of baseball, relatively few home runs were hit and it was important to use players' speed to hit triples to score runs. In recent years, speed has played much less of a factor in scoring runs, and the home run is an important offensive weapon. So the association structure in this scatterplot tells us how the game of baseball has changed in the last 100 years.

Case Study 4-2: Runs and Offensive Statistics

Topics Covered: Scatterplots and looking for association, correlation   Let's talk more about baseball offensive statistics:

- Since the objective of a baseball team is to score more runs than its opponent, the most important offensive statistic is runs.

- The value of any offensive statistic depends on its relationship with runs.

- We are interested in finding the offensive statistic that has the strongest association with runs.

Here is a list of the basic offensive stats.

H   2B   3B   HR   RBI   AVG   TB   SLG   OBP

How strong is the relationship of these statistics with runs scored (R) by a team? We make some preliminary observations based on our knowledge of baseball.

- Clearly, RBI will have a strong positive relationship with runs, since RBI counts the number of runs that are batted in by a team.

- Triples (3B), in contrast, seem to have a relatively weak relationship with runs. Teams that get a lot of triples don't necessarily score more runs. This is true since triples are a relatively rare event and a large number of triples might just be a reflection of the team's speed or the configuration of the ballpark.

- Home runs (HR) and doubles (2B) have stronger relationships with runs since they occur more frequently than triples and do typically result in runs. (Home runs actually cause at least one run to be scored.)

- TB (total bases) and SLG (slugging percentage) are essentially the same stat (you divide TB by AB to get SLG), so they both have about the same relationship with runs.

- AVG is probably less associated than OBP or SLG with runs. Unlike SLG, AVG doesn't give added weight to extra base hits (doubles, triples, and home runs) over singles; and unlike OBP, AVG doesn't reflect walks or HBP, which both contribute to runs scored by a team.

From this discussion, we get the following list of the statistics ranked with respect to our expected strength of the association with runs scored.

Statistic	Rank
RBI	High association with runs
OBP	
SLG, TB	
AVG	
HR	
2B	
3B	Low association with runs

At this point, we are thinking about association by means of the pattern that we see in the scatterplot. We can summarize these visual associations by means of correlations.

We return to the team offensive stats from the 2000 season discussed previously. We're interested in the offensive stats that are most associated with runs scored. Look at the scatterplot matrix in Figure 4.5 and focus on the plots in the first column—these are the ones that plot runs (R) against OBP, SLG, 2B, 3B, HR.

Some comments:

- Runs are strongly positively associated with OBP and SLG, although the relationship with OBP looks a bit stronger.

- Runs are somewhat positively associated with HR—the association is less strong than the association with OBP and SLG.

- The relationship between Runs and 2B, and Runs and 3B, looks weak.

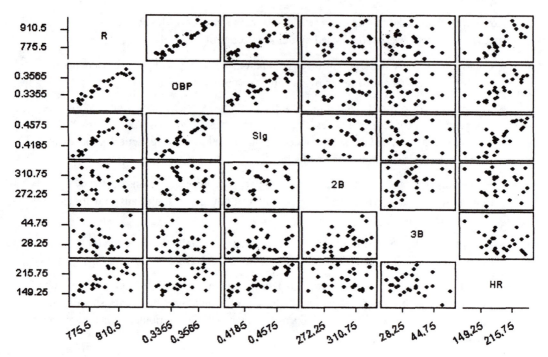

FIGURE 4.5
Scatterplot matrix of different offensive statistics for 30 Major League teams in the 2000 season.

We can support these comments by computing correlations. The matrix in Table 4.3 gives correlations between the different offensive team statistics.

TABLE 4.3
Correlation matrix between a number of different offensive team statistics.

	R	H	2B	3B	HR	RBI	AVG	TB	SLG
H	0.747								
2B	0.238	0.535							
3B	0.061	0.280	0.430						
HR	0.644	0.234	−0.050	−0.344					
RBI	0.996	0.734	0.248	0.062	0.678				
AVG	0.793	0.980	0.461	0.260	0.294	0.778			
TB	0.869	0.797	0.446	0.097	0.748	0.886	0.813		
SLG	0.866	0.699	0.347	0.051	0.824	0.886	0.750	0.979	
OBP	0.934	0.687	0.188	0.032	0.609	0.927	0.762	0.804	0.826

Using these correlation values, we can rank the most valuable offensive stats:

Correlation with runs

OBP	.93
SLG	.87
AVG	.79
HR	.64
2B	.23
3B	.08

Of the three "averages," AVG has the weakest correlation with runs scored, while OBP is the best with SLG nearly as good. Of the three count variables, HR is by far the most strongly correlated with runs scored, but it lags well behind the three averages. Doubles (2B) and 3B are far inferior to HR, with 3B virtually useless.

Case Study 4-3: Most Valuable Hitting Statistics

Topics Covered: Linear regression, prediction   There is a group of baseball fans who are members of SABR, the Society of American Baseball Research. These people are interested in the history of the game and write many books and articles about the game. Sabermetrics is the mathematical and statistical study of baseball records. We focus on one of the most important topics to sabermetricians: What is the most valuable hitting statistic?

Who is the better hitter, Jose Cruz or Mark McLemore? This can be a difficult question to answer since Cruz and McLemore are different types of hitters. Cruz is a power hitter who hits many doubles and home runs. McLemore, in contrast, is more of a singles hitter who gets on base with his speed. Your decision about who is the better hitter depends partly on how you value power hitting relative to on-base ability. We want to find statistics that can distinguish the hitting accomplishments of Cruz and McLemore.

Let's review the basic hitting statistics recognized by Major League Baseball.

- AVG, the batting average, is the most commonly quoted hitting statistic of Major League Baseball. The player who wins the batting crown is the one with the highest batting average.

- OBP, the on-base percentage.

- SLG, the slugging percentage.

Starting in the 1960s, a large number of new statistics have been introduced as "improvements" to the basic three statistics:

- OPS, the sum of OBP and SLG.

- Runs created (RC) introduced by Bill James. It is defined as

$$RC = (H + BB)TB/(AB + BB).$$

- Total average (TA). It is like SLG, but it is the ratio of bases to the number of outs:

$$TA = (TB + BB)/(AB - H).$$

- Batter's runs average (BRA), the product of OBP and SLG.

How do we evaluate all of these hitting statistics?

First, we have to understand that the goal of hitting is to create runs, and a single player can't create a run (unless he hits a home run). Teams create runs and so we have to look at team data to understand the usefulness of these various statistics.

Let's consider team stats for the 2000 American League teams displayed in Table 4.4. (We present data for only one league in this case study to make it easier to present the methodology for evaluating a hitting statistic.)

TABLE 4.4
Offensive statistics for the 2000 American League teams.

Team	R	G	R/G	AVG	OBP	SLG	OPS	RC	TA	BRA
Anaheim	864	162	5.33	0.280	0.352	0.472	0.824	930	0.805	0.166
Baltimore	794	162	4.90	0.272	0.341	0.435	0.776	817	0.735	0.148
Boston	792	162	4.89	0.267	0.341	0.423	0.764	808	0.725	0.144
Chicago AL	978	162	6.04	0.286	0.356	0.470	0.826	939	0.805	0.167
Cleveland	950	162	5.86	0.288	0.367	0.470	0.837	975	0.830	0.172
Detroit	823	162	5.08	0.275	0.343	0.438	0.781	843	0.741	0.150
Kansas City	879	162	5.43	0.288	0.348	0.425	0.773	842	0.723	0.148
Minnesota	748	162	4.62	0.270	0.337	0.407	0.744	768	0.693	0.137
New York AL	871	161	5.41	0.277	0.354	0.450	0.804	878	0.780	0.159
Oakland	947	161	5.88	0.270	0.360	0.458	0.818	908	0.812	0.165
Seattle	907	162	5.60	0.269	0.361	0.442	0.803	873	0.798	0.160
Tampa Bay	733	161	4.55	0.257	0.329	0.399	0.728	715	0.674	0.131
Texas	848	162	5.23	0.283	0.352	0.446	0.798	882	0.766	0.157
Toronto	861	162	5.31	0.275	0.341	0.469	0.810	897	0.775	0.160

The important statistic for a team is the number of runs scored. If we divide this number by the number of games, we get the RUNS/GAME statistic (or R/G). For example, Chicago scored 978 runs in 162 games, so

$$R/G = 978/162 = 6.04.$$

We are interested in seeing the effectiveness of different hitting statistics in predicting R/G. Let's start with the traditional AVG measure. How well can we predict a team's runs per game (R/G) using AVG? If we do a scatterplot of R/G against AVG (shown in Figure 4.6) we see a positive association—teams that hit for high averages tend to score more runs.

A least-squares fit to these data gives the relationship

$$R/G = 32.6 \, AVG - 3.69.$$

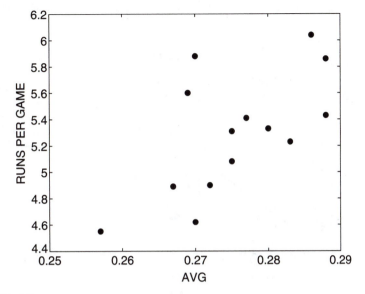

FIGURE 4.6
Scatterplot of batting average and runs scored per game for 2000 American League teams.

Is this a good fit? In other words, how close are the points in the scatterplot to the fitted line?

We evaluate the goodness of fit of this prediction equation in several steps. The calculations are summarized in the following table.

1. First, we find the predicted values of R/G for each team. For example, we see that Anaheim had a batting average of .280. We would predict its R/G to be R/G = 32.6(.280) − 3.69 = 5.44. This value is placed in the Predicted R/G column.

2. The residual is the difference between the actual R/G for a team and its predicted R/G. In the case of Anaheim, it actually scored an average of 5.33 runs per game and we predicted 5.44, so

$$\text{Residual} = \text{Actual R/G} - \text{Predicted R/G} = 5.33 - 5.44 = -0.11$$

In Figure 4.7, we have drawn the least-squares line on the scatterplot. The vertical lines drawn from each point represent the residuals. (The least-squares line minimizes the sum of the squared residuals.) We have identified two teams with unusually large residuals. Oakland and Minnesota both had a season batting average of .270. However, Oakland R/G in 2000 was much greater than the R/G predicted by using batting average and the residual is a large positive value. In contrast, Minnesota scored a very small number of runs given its batting average and has a large negative residual.

3. We summarize the sizes of these residuals by use of a Root Mean Square Error (RMSE) criterion. Table 4.5 illustrates the calculation of the RMSE. We first square all the residuals and put the answers in the Sq Residual column.

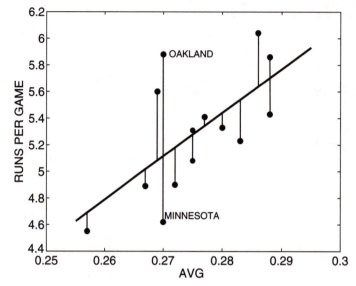

FIGURE 4.7
Scatterplot of batting average and runs scored per game for 2000 American League teams, with least-squares fit and residuals displayed.

TABLE 4.5
Calculation of the Root Mean Square Error criterion for the least-squares fit with AVG as the predicting variable.

Team	AVG	R/G	Predicted R/G	Residual	Sq Residual
Anaheim	0.280	5.33	5.44	−0.11	0.0117
Baltimore	0.272	4.90	5.18	−0.28	0.0768
Boston	0.267	4.89	5.01	−0.12	0.0154
Chicago$_{AL}$	0.286	6.04	5.63	0.41	0.1652
Cleveland	0.288	5.86	5.70	0.16	0.0260
Detroit	0.275	5.08	5.28	−0.20	0.0380
Kansas City	0.288	5.43	5.70	−0.27	0.0723
Minnesota	0.270	4.62	5.11	−0.49	0.2421
NewYork$_{AL}$	0.277	5.41	5.34	0.07	0.0049
Oakland	0.270	5.88	5.11	0.77	0.5898
Seattle	0.269	5.60	5.08	0.52	0.2710
Tampa Bay	0.257	4.55	4.69	−0.14	0.0191
Texas	0.283	5.23	5.54	−0.31	0.0935
Toronto	0.275	5.31	5.28	0.03	0.0012
			TOTAL	1.6270
			MEAN	0.116
			RMSE	0.341

The Sum of Squared Errors is the sum of values in the Sq Residual column. Here it is 1.627. The mean of the Squared Residuals, called Mean Square Error or MSE, is

$$MSE = 1.627/14 = 0.116.$$

The Root Mean Square Error, RMSE, is the square root of MSE:

$$RMSE = \sqrt{0.116} = 0.341.$$

The RMSE is a measure of the size of the residuals from the model that uses AVG to predict R/G.

Here is a nice interpretation of RMSE: Generally, if you graph all of the residuals, you will find them approximately normally distributed with mean 0 and standard deviation RMSE. Since we have a normal distribution, we can apply the 68–95–99.7 rule and say that

68% (roughly $\frac{2}{3}$) of the residuals fall between $-$RMSE and $+$RMSE

To illustrate this interpretation, suppose we use AVG to predict R/G for the American League teams. We found that RMSE = 0.341. This means if you use AVG to predict R/G, then roughly $\frac{2}{3}$ (about 9) of the residuals will fall between -0.341 and $+0.341$.

How Good Are Other Batting Measures?

Now let's try using the batting statistic OBP to predict R/G. The least-squares line is

$$R/G = 38.8(OBP) - 8.24.$$

As we did above, we find the predicted value of R/G for each team and compute the residuals and squared residuals and put the results in Table 4.6.

The sum of squared residuals is equal to .5344 and the RMSE is equal to

$$RMSE = \sqrt{0.5344/14} = .195.$$

Let's summarize what we learned. The RMSE is a measure of the size of the residuals in our prediction. When we used AVG to predict Runs/Game, the RMSE was equal to .341; when we used OBP, the RMSE is equal to .195. This tells us that the residuals are generally much larger using AVG instead of using OBP as a predictor. This means that OBP is a much better predictor of Runs Scored than AVG using the RMSE criterion. (In general, the smaller the RMSE, the closer the predictions.)

A Prediction Contest

We used each of our seven batting statistics to predict runs per game for teams for our American League data. Table 4.7 displays the corresponding values of RMSE.

We see that OBP, OPS (OBP + SLG), TA, RC, and BRA (OBP $\times$ SLG) are all pretty close at the top, SLG is a bit behind, and AVG is pulling up the rear.

TABLE 4.6
Calculation of the Root Mean Square Error criterion for the least-squares fit with OBP as the predicting variable.

Team	OBP	R/G	Predicted R/G	Residual	Sq Residual
Anaheim	0.352	5.33	5.42	−0.09	0.0077
Baltimore	0.341	4.90	4.99	−0.09	0.0082
Boston	0.341	4.89	4.99	−0.10	0.0102
Chicago AL	0.356	6.04	5.57	0.47	0.2183
Cleveland	0.367	5.86	6.00	−0.14	0.0195
Detroit	0.343	5.08	5.07	0.01	0.0001
Kansas City	0.348	5.43	5.26	0.17	0.0281
Minnesota	0.337	4.62	4.84	−0.22	0.0465
New York AL	0.354	5.41	5.50	−0.09	0.0073
Oakland	0.360	5.88	5.73	0.15	0.0231
Seattle	0.361	5.60	5.77	−0.17	0.0278
Tampa Bay	0.329	4.55	4.53	0.02	0.0006
Texas	0.352	5.23	5.42	−0.19	0.0352
Toronto	0.341	5.31	4.99	0.32	0.1019
				TOTAL	0.5344
				RMSE	0.1950

TABLE 4.7
Values of the root mean squared criterion using each of seven batting statistics to predict the runs scored per game for the 2000 American League team data.

Statistic	RMSE
AVG	.341
OBP	.195
SLG	.265
OPS	.205
TA	.202
RC	.204
BRA	.195

Of course, we just looked at 2000 data. Is it possible that another batting statistic would do better a different year? Albert and Bennett (2001) looked at all years from 1952 to 1999. Each year, they used each statistic to predict runs per game for the teams data, and found RMSE for each statistic. Albert and Bennett's general conclusions were:

- AVG performs terribly in predicting runs scored. This measure should be thrown out (but it probably won't be).

- Both SLG and OBP are relatively mediocre measures. (It is interesting that OBP did so well in 2000. It is not clear why.)

- The "brainstorming" statistics TA, BRA, RC, OPS are all good measures and they are pretty close. You can't say for sure that one statistic in this group is better than another in the group.

Remember our motivation for finding the best hitting statistic? We want to use a good statistic to compare the relative worth of two hitters.

To illustrate the use of good hitting statistics to compare players, let's compare Mark McLemore and Jose Cruz in the 2001 baseball season. Table 4.8 gives the batting average and the seven alternative batting measures for both McLemore and Cruz for this season.

TABLE 4.8
Batting measures for Mark McLemore and Jose Cruz for the 2001 baseball season.

Batters	AVG	SLG	OBP	OPS	RC	TA	BRA
McLemore	0.286	0.406	0.384	0.790	64.6	0.805	0.156
Cruz	0.274	0.530	0.326	0.856	99.9	0.838	0.173

Note from the table that, although McLemore has a higher AVG and OBP, Cruz was substantially better than McLemore using each of the five "new" batting statistics. Cruz clearly had a much better hitting year in 2001.

Case Study 4-4: Creating a New Measure of Offensive Performance Using Multiple Regression

Topics Covered: Multiple linear regression, root mean square error, least-squares criterion
In the earlier case studies, we introduced a number of batting measures, some old and some new, for evaluating the value of a hitter. We used the 2000 American League team data to evaluate these different measures. Specifically, we used each measure to predict the average runs per game for the 14 AL teams, and compared the measures by computing the average size of a residual (RMSE). We found that many of the modern batting measures, such as OPS and RC, did substantially better compared to the traditional measures AVG and SLG for predicting run production.

There is actually a straightforward way of finding a "best" measure of hitting performance using the tool of multiple linear regression.

The basic batting counts are the numbers of singles, doubles, triples, and home runs (1B, 2B, 3B, and HR, respectively), and the number of walks (BB) and hit-by-pitch (HBP). The goal is to combine these different counts in some way to obtain an accurate prediction of the runs scored per game.

Many of the standard batting measures combine these batting counts in a linear way. For example, the batting average AVG can be expressed as

$$AVG = (1/AB)1B + (1/AB)2B + (1/AB)3B + (1/AB)HR,$$

where singles, doubles, triples, and home runs are given equal weights. The slugging percentage SLG is also a linear measure of the form

$$\text{SLG} = (1/\text{AB})\,1\text{B} + (2/\text{AB})\,2\text{B} + (3/\text{AB})\,3\text{B} + (4/\text{AB})\,\text{HR},$$

where a hit is weighted by the number of bases. The on-base percentage, OBP, weights all on-base events (hits, walks, and hit-by-pitch) equally:

$$\text{OBP} = (1/\text{PA})\,\text{BB} + (1/\text{PA})\,\text{HPB} + (1/\text{PA})\,1\text{B} + (1/\text{PA})\,2\text{B} + (1/\text{PA})\,3\text{B} + (1/\text{PA})\,\text{HR},$$

where $\text{PA} = \text{AB} + \text{BB} + \text{HBP}$ is the number of plate appearances.

Suppose that we consider an alternative batting measure, called OPTAVG (for optimal average), that combines all of the batting events in a linear way with arbitrary weights:

$$\text{OPTAVG} = w_0 + w_1 \times 1\text{B} + w_2 \times 2\text{B} + w_3 \times 3\text{B} + w_4 \times \text{HR} + w_5 \times \text{BB} + w_6 \times \text{HBP}.$$

Can we find values of the weights $w_0, \ldots, w_6$ that give a "best" batting measure?

Fortunately, there is an easy way to find optimal weights for this batting measure using a least-squares criterion. As in the earlier case study, suppose that we have team hitting data. For each team, we observe the runs scored per game (R/G) and the counts of the six batting events (1B, 2B, 3B, HR, BB, HBP). The goal is to predict R/G based on the linear measure OPTAVG. Using the least-squares criterion, we wish to find values of the weights $w_0, \ldots, w_6$ that minimize the sum of squared residuals

$$\text{Sum of } (\text{OPTAVG} - \text{R/G})^2.$$

Values of these weights, called the least-squares estimates, are easily available using any standard statistical computing package.

We apply this method to estimate R/G for the 30 Major League Teams in 2000. Here is output from MINITAB:

```
The regression equation is
R/G = - 4.20 + 0.00481 1B + 0.00277 2B + 0.00831 3B + 0.00826 HR
            + 0.00295 BB + 0.00263 HBP
```

Before we interpret the weights of this measure, let's explain why this is the best linear batting measure. Among all batting statistics that combine the different batting events (single, double, etc.) in a linear way, this measure will give the smallest value of RMSE, which is an average size of a residual when this measure is used as a predictor. Since this "best linear" measure has the smallest RMSE, it will have a smaller RMSE than other linear measures such as AVG, SLG, OBP, and OPS.

To demonstrate this fact, we also tried using AVG, SLG, OBP, OPS, and RC, each alone, as predictors for this 2000 Major League team dataset. Table 4.9 gives values of RMSE for all 30 Major League teams in 2000 using the best linear and the more familiar measures.

We see that the best linear measure OPTAVG has an RMSE value, 0.126, that appears to be much smaller than its nearest competitor RC (0.162) and the popular OPS statistic (0.175). But this difference is a bit deceptive since we found the best weights using the

TABLE 4.9
Values of the root mean squared criterion using the best linear and other measures for the 2000 Major League team data.

Measure	OPTAVG	AVG	SLG	OBP	OPS	RC
RMSE	0.126	0.290	0.236	0.163	0.175	0.162

same 2000 dataset—it is not clear that this best measure will also be good in predicting the runs scored per game for team data for the 2001 year.

The weights of the best linear measure OPTAVG tell about the worth of each type of batting event. When we compute a batting average, we give each type of hit the same weight and we ignore walks and hit-by-pitches. A slugging percentage weights each hit by the number of bases produced. What does our best linear measure do for weights? Table 4.10 shows the values of the weights for each base event, and, to make the comparison easy, it standardizes the weights by dividing each by the weight of a single.

TABLE 4.10
Weights of batting events for the best linear measure OPTAVG.

Event	Single	Double	Triple	Home Run	Walk	Hit by Pitch
Weight	0.00481	0.00277	0.00831	0.00826	0.00295	0.00263
Wt/Wt(Single)	1	0.58	1.73	1.72	0.61	0.54

Some of the values of these weights are a bit surprising. The weight for a double is actually smaller than the weight given to a single! A triple is weighted less than two times a single and a triple and home run are given roughly the same weights. How can we explain these weights?

The first comment is that we are using a relatively small amount of data and these unusual weight values are a consequence of the fact that we are fitting a relatively complicated model to a small dataset. We could get more reasonable estimates at these weights by using, say, 10 years of team data instead of just one year. Albert and Bennett (2001, Chapter 5) do exactly that and get more reasonable weights for this linear batting measure.

However, some of these weights do make sense. It is well known by quantitative baseball people that a triple is not three times the worth of a single—its actual worth is closer to two times a single, which is the value in this table. Similarly, it makes sense that the home run is not worth four times a single. Walks, hit-by-pitches, and singles should have weights of a similar size since all these events get the batter to first base. But the single is a more valuable event than a walk or hit-by-pitch since the runners are typically advanced more than one base. In other words, the larger weight for a single reflects the fact that the single is more effective than a walk or hit-by-pitch in producing runs.

Case Study 4-5: How Important Is a Run?

Topics Covered: Nonlinear regression, transformation, residuals   In this chapter, we have discussed how it is important for a baseball team to score runs and we evaluated the goodness of different batting measures by their relationship with runs scored. But of course the objective of a baseball team is not solely to score runs—it wins games by scoring more runs than their opponent.

That raises the interesting question, what is the importance of a single run toward the goal of winning a baseball game? If a player is responsible for scoring, say, 20 runs, then how many wins for his team has he contributed? Bill James discovered a special relationship between the number of wins (W) and losses (L) for a baseball team and the number of runs scored (R) and number of runs allowed (RA). He called this relationship "The Pythagorean Method." This result says that the ratio between a team's wins and losses is approximately equal to the square of the ratio of runs scored to runs allowed. That is, approximately,

$$\frac{W}{L} = \left(\frac{R}{RA}\right)^2.$$

If we take logs of both sides, we get the equivalent relationship

$$\log\left(\frac{W}{L}\right) = 2\log\left(\frac{R}{RA}\right).$$

(We take logs to convert a nonlinear equation in the runs ratio to a linear equation.) Can we demonstrate that the Pythagorean Method gives a good description of the relationship between the win/loss pattern and the runs scored/allowed for current Major League Baseball teams?

To answer this question, we look at the relevant team statistics (wins, losses, runs scored, and runs allowed) for the 30 teams for the 2000 season. Looking at Table 4.11, we see that if a team has a winning record, then it generally scores more runs than its opponent. But there is one interesting exception—Toronto had a win/loss record of 83-79 but actually allowed 47 more runs this season than they scored. (We suppose that Toronto won a lot of close games in 2000.)

To look for the Pythagorean relationship, we compute log(W/L) and log(R/RA) for all teams and construct a scatterplot of the two quantities in Figure 4.8. We see a linear positive association in this graph, indicating that there is indeed a linear association between log(W/L) and log(R/RA).

Next we want to fit a "best line" to this graph. It seems natural to restrict this line to pass through one point. If a team scores the same number of runs against its opponents (R = RA), then we expect the team to win half of its games (W = L). In other words, the point (log(R/RA), log(W/L)) = (0, 0) should fall on the line. With this restriction, we look at line fits of the form

$$\log(W/L) = k \log(R/RA).$$

TABLE 4.11

Team statistics for the Major League teams in the 2000 season.

Team	W	L	R	RA	Team	W	L	R	RA
Anaheim	82	80	864	869	Milwaukee	73	89	740	826
Arizona	85	77	792	754	Minnesota	69	93	748	880
Atlanta	95	67	810	714	Montreal	67	95	738	902
Baltimore	74	88	794	913	New York AL	87	74	871	814
Boston	85	77	792	745	New York NL	94	68	807	738
Chicago AL	95	67	978	839	Oakland	91	70	947	813
Chicago NL	65	97	764	904	Philadelphia	65	97	708	830
Cincinnati	85	77	825	765	Pittsburgh	69	93	793	888
Cleveland	90	72	950	816	San Diego	76	86	752	815
Colorado	82	80	968	897	San Francisco	97	65	925	747
Detroit	79	83	823	827	Seattle	91	71	907	780
Florida	79	82	731	797	St. Louis	95	67	887	771
Houston	72	90	938	944	Tampa Bay	69	92	733	842
Kansas City	77	85	879	930	Texas	71	91	848	974
Los Angeles	86	76	798	729	Toronto	83	79	861	908

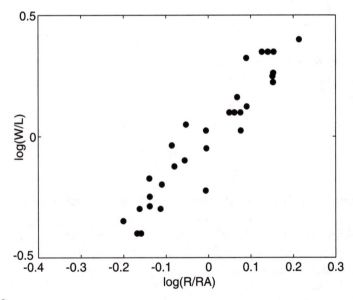

FIGURE 4.8

Scatterplot of log runs ratio against log of ratio of wins to losses for Major League team data from the 2000 season.

We choose *k* by using a least-squares criterion. It turns out that the sum of squared residuals is minimized when $k = 1.91$. Figure 4.9 shows this best line on the scatterplot and a display of the corresponding residuals. We do not see any linear trend or any other pattern in the residual plot, so it appears that our fit is satisfactory.

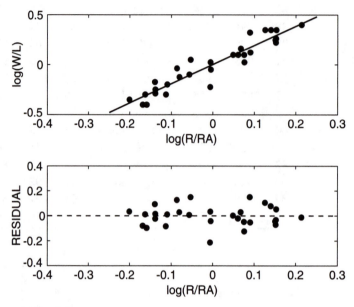

FIGURE 4.9
Least-squares fit (top) and residual plot (bottom) for (R/RA, W/L) data.

So based on our analysis, we arrive at the relationship

$$\frac{W}{L} = \left(\frac{R}{RA} \right)^{1.91}$$

which is pretty close to James's Pythagorean relationship, which uses the power of 2. How useful is this rule in predicting a team's win numbers? To check the accuracy of this relationship in prediction, Table 4.12 gives the actual number of wins, the predicted number of wins (using the above model) and the residual (actual − predicted). Figure 4.10 displays a stemplot of the absolute residuals.

We see from the stemplot that 24 of the 30 residuals sizes are smaller than 4. This indicates that for 80% of the teams, we can predict the number of wins to within four games using this formula.

Case Study 4-6: Baseball Players Regress to the Mean

Topics Covered: Least-squares regression, prediction   Here we examine least-squares regression. Specifically, we describe the "regression effect" that is generally unknown to

TABLE 4.12
Number of wins, predicted number of wins, and residuals using James's Pythagorean relationship.

Team	W	Predicted	Residual	Team	W	Predicted	Residual
Anaheim	82	80.6	1.4	Milwaukee	73	72.5	0.5
Arizona	85	84.8	0.2	Minnesota	69	68.5	0.5
Atlanta	95	90.7	4.3	Montreal	67	65.6	1.4
Baltimore	74	70.2	3.8	New York$_{AL}$	87	85.7	1.3
Boston	85	85.7	−0.7	New York$_{NL}$	94	87.9	6.1
Chicago$_{AL}$	95	92.8	2.2	Oakland	91	92.2	−1.2
Chicago$_{NL}$	65	68.1	−3.1	Philadelphia	65	68.8	−3.8
Cincinnati	85	86.8	−1.8	Pittsburgh	69	72.3	−3.3
Cleveland	90	92.7	−2.7	San Diego	76	74.8	1.2
Colorado	82	86.9	−4.9	San Francisco	97	97.3	−0.3
Detroit	79	80.6	−1.6	Seattle	91	92.6	−1.6
Florida	79	73.9	5.1	St. Louis	95	91.8	3.2
Houston	72	80.5	−8.5	Tampa Bay	69	69.9	−0.9
Kansas City	77	76.6	0.4	Texas	71	70.3	0.7
Los Angeles	86	88	−2	Toronto	83	76.9	6.1

```
        ABSOLUTE RESIDUALS
     0 | 23455779
     1 | 22344668
     2 | 027
     3 | 12388
     4 | 39
     5 | 1
     6 | 11
     7 |
     8 | 5
```

FIGURE 4.10
Stemplot of the absolute residuals from the fit using James's Pythagorean relationship.

most people in baseball. If you look up the word "regress" in the dictionary, it will tell you the word means to "go back." We will see that there is a general tendency for a player's baseball stats from one year to the next to go back, or regress to the mean.

Many of you have heard about the so-called "sophomore slumps" in sports. This happens when someone does well in his/her rookie year and then slumps in the sophomore year. This tendency is commonly discussed among baseball people. Here we see that there is a natural tendency for best performing rookies to slump their sophomore years.

Let's look at the OBP data for eight players from the Yankees and eight players from the Mets. Table 4.13 gives the 2000 and 1999 OBPs for these players. In addition, in the column labeled "Improvement," we show how many OBP points a player improved from

TABLE 4.13
1999 and 2000 on-base percentages for 16 players from the Mets and the Yankees.

	OBP_{1999}	OBP_{2000}	Improvement
Jeter	438	416	−22
O'Neill	353	336	−17
Justice	413	377	−36
Williams	435	391	−44
Brosius	307	299	−8
Knoblauch	393	366	−27
Martinez	341	328	−13
Posada	341	417	76
Piazza	361	398	37
Alfonzo	385	425	40
Bordick	334	341	7
Harris	330	317	−13
Ventura	379	338	−41
Zeile	354	356	2
Agbayani	363	391	28
Hamilton	386	358	−28

1999 to 2000; if this improvement is negative, this means that the player had a lower OBP in 2000.

Suppose that some player has a 400 on-base percentage in 1999. What do you predict his OBP value to be in 2000 (assuming that you don't know his 2000 data yet)? If you don't know anything about the hitter, it would seem reasonable to predict his 2000 OBP also to be 400. Is this the best prediction? To investigate this, we first plot the players' 2000 OBP against the players' 1999 values in Figure 4.11.

We see a relatively strong positive association in this graph. Players that were good in 1999 in terms of getting on-base (like Derek Jeter) tended to be good in 2000. Likewise, players with small OBP values in 1999 (Scott Brosius comes to mind) tended to be poor also in 2000.

The least-squares line (plotted on the scatterplot) is

$$OBP\ 2000 = 0.615(OBP\ 1999) + 138.$$

Let's illustrate using this least-squares line to predict a player's 2000 OBP. Let us select Bernie Williams of the Yankees. His 1999 OBP value was 435. Using the line, we would predict his 2000 OBP value to be

$$OBP\ 2000 = 0.615(435) + 138 = 406.$$

Let's interpret this prediction:

1. In 1999, Williams' OBP was 435. Since the mean OBP (among these 16 players) was 370, Bernie was 65 points above average in 1999.

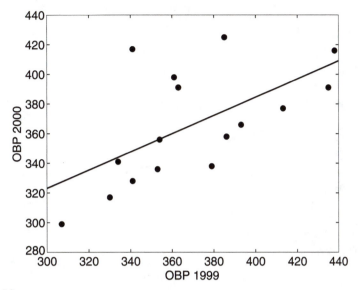

FIGURE 4.11
Scatterplot of 1999 and 2000 on-base percentages for 18 players. A least-squares line is shown on the graph.

2. We would predict Williams' 2000 OBP to be 406. Since the mean OBP in 2000 was 366, our prediction is 40 points above average.

So his performance in 1999 was 65 points above average, and we predict his 2000 performance to be 40 points above average. In other words,

> We predict that his 2000 OBP will be closer to the mean than his 1999 OBP.

This demonstrates a general tendency for a player's OBP to regress toward the mean.

Let's explain this regression effect a different way. Recall that we defined a player's Improvement as

$$\text{Improvement} = (\text{OBP } 2000) - (\text{OBP } 1999).$$

For example, Derek Jeter's improvement would be $416 - 438 = -22$. His OBP dropped 22 points from 1999 to 2000. Looking at the data, we see that

- the best improvement was Posada at $+76$,

- the most negative improvement was Bernie Williams at -44.

Suppose that we compute the improvement for all players and graph the Improvement against the 1999 OBP. The scatterplot is shown in Figure 4.12. We see that good players in 1999 (with large OBPs) tend to have negative improvements, and poor players in 1999 (with small OBPs) tend to improve in 2000. There is a negative association between the improvement and the first-year OBP.

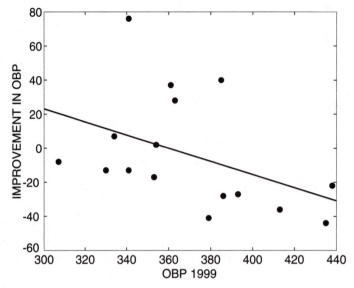

FIGURE 4.12

Scatterplot of 1999 on-base percentage and improvement in OBP from 1999 to 2000 for 18 players. A least-squares fit is drawn on the graph.

The line in the graph is a least-squares fit. The correlation $r = -0.43$ and the least-squares line is

$$\text{Improvement} = 139 - 0.39(\text{OBP } 1999).$$

This "regression effect" actually is very common. If you take statistics for a group of players for two consecutive years, you will find that a player's improvement is negatively associated with his first year performance. Players' statistics (from one year to the next) tend to move toward the average value.

Case Study 4-7: The 2000 Dinger Drop-off

Topics Covered: Scatterplots, least-squares line, correlation, comparison of batches by stemplots, five-number summaries   In this study we examine home runs in 2000. The year 2000 was a big year in terms of the number of home runs hit. There were a record number of home runs, despite the injuries of the big sluggers, Mark McGwire and Ken Griffey, Jr.

There were a couple of interesting features of this season with respect to home runs.

1. The total number of home runs was high—we compare the 2000 home run numbers with the numbers in the 1999 season.

2. A *Sports Illustrated* article, published October 2000, titled "Dinger Drop-Off," notes the change in the pattern of home run hitting between the first half and second half of the 2000 season. We will look at this change by using a scatterplot.

We begin by collecting the home run totals for all teams in the 1999 and 2000 baseball seasons. To compare these two groups of team totals, Figure 4.13 shows back-to-back stemplots.

1999 HOMERUNS		2000 HOMERUNS
5	10	
	11	6
8	12	
	13	
5	14	4
831	15	07
85321	16	01278
61	17	377899
9871	18	34
743	19	88
993	20	05
622	21	16
3	22	16
50	23	569
4	24	49

FIGURE 4.13
Back-to-back stemplots of the team home run totals in the 1999 and 2000 baseball seasons.

Both distributions of home run totals look a bit right-skewed with centers in the 160–170 range. To make a more precise comparison of these two groups, we compute five-number summaries. For each group, there are 30 home run totals. The position of the median M is 15.5 (halfway between the 15th and 16th observations) and the position of the quartile is 8. The lower quartile is the 8th smallest value and the upper quartile is the 8th largest value.

From the stemplots, we find that the five-number summaries are:

1999 home run team totals: (105, 162, 187.5, 209, 244),
2000 home run team totals: (116, 167, 179, 216, 249).

Comparing medians (187.5 for 1999 and 179 for 2000), we see that the typical team hit eight more home runs in 1999 than in 2000. Wait a second ... didn't I say that there were more home runs hit in 2000?

Yes, there were more home runs hit in 2000. We can compute the **home run totals** by looking at the sample means. For the 2000 data, $\bar{x} = 189.77$. So the total number of home runs is equal to $189.77 \times 30 = 5{,}693$. For the 1999 data, $\bar{x} = 184.27$. Here the total number of home runs is $184.27 \times 30 = 5{,}528$. So there were $5693 - 5528 = 165$ more home runs hit in 2000.

To summarize, the medians are useful in comparing "typical home run hitting teams," and the means are helpful in computing home run totals.

Next, we look at the relationship between the number of home runs hit by a team *before* the All-Star game and the home run production of the team *after* the All-Star game. This is an example of **breakdown data** that is currently very common in baseball statistics.

The book web site (described in Appendix 2) gives the number of home runs hit before (PRE_HR) and after (POST_HR) the All-Star game for all 30 teams. Here are several comments about these data.

1. Note that most teams hit more home runs before the All-Star game than after the All-Star game. But this is not a fair comparison, since more games were played before the All-Star game than after.

2. We can confirm the above comment by looking at the number of at-bats in the table. For example, St. Louis had more home runs in the first half (149) than the second half (86), but they also had more at-bats in the first half (2946) than the second half (2532).

3. We can adjust the home run totals for the different number of at-bats by computing **home run rates**:

$$\text{home run rate} = \text{HR/AB}.$$

We look for a relationship between a team's pre-All-Star home run rate and it's post-All-Star home run rate by constructing a scatterplot in Figure 4.14. A 45-degree line is plotted on the graph—this line will be used to help in our interpretation.

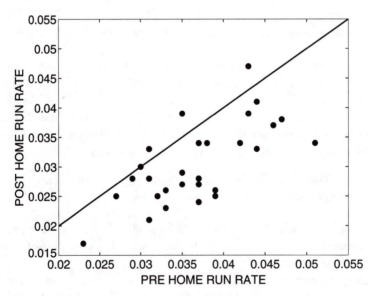

FIGURE 4.14
Scatterplot of team home run rates before and after the All-Star break for the 2000 season.

1. We see a positive association between the two home run rates. This means that teams that hit for a high rate before the break tended to hit for a high rate after the break;

likewise teams that had low home run rates in the first half of the season tended to have low rates in the second half.

2. We can guess at the correlation value—a value like $r = 0.5$ would be a reasonable guess. (Actually the value of r is .68.)

3. A line is plotted on the graph. This line represents values where the pre-home run rate is equal to the post-home run rate. One team (Pittsburgh) had equal home run rates for both halves—that corresponds to the point on the line.

4. Counting, we see that 26 points are *under* the line-these teams had a lower home run rate after the break. The three points *above* the line correspond to teams that had a higher home run rate after the break.

Since 26 of the 30 points are under the line, we see a general tendency for teams to hit for a smaller rate of home runs after the break. Why?

Here are some possible explanations for the home run drop:

1. Pitching improved during the season.
2. Players got fatigued over the season.
3. The big home run hitters such as Mark McGwire were injured.
4. The composition of the baseball changed (this is mentioned in the article in *Sports Illustrated*).

Exercises

Leadoff Exercise. Table 4.14 shows career on-base percentage (OBP) and slugging percentage (SLG) for Rickey Henderson for the first 23 seasons of his career.

TABLE 4.14
On-base and slugging percentages for Rickey Henderson for his first 23 seasons of his career.

Age	OBP	SLG	Age	OBP	SLG
20	.338	.336	32	.400	.423
21	.420	.399	33	.426	.457
22	.408	.437	34	.432	.474
23	.398	.382	35	.411	.365
24	.414	.421	36	.407	.447
25	.399	.458	37	.410	.344
6	.419	.516	38	.400	.342
27	.358	.469	39	.376	.347
28	.423	.497	40	.423	.466
29	.394	.399	41	.368	.305
30	.411	.399	42	.366	.351
31	.439	.577			

(a) Construct a scatterplot of Rickey's OBP and SLG values where OBP is the horizontal variable and SLG is the vertical variable.

(b) There are five unusual points on the left side of the plot that don't follow the general pattern. This is where Rickey had a small slugging percentage. Find the ages that correspond to these unusual points.

(c) Circle the point that corresponds to Rickey's best season with respect to both OBP and SLG. What was Rickey's age this particular year?

(d) If you ignore the five unusual points, what is the general pattern in the scatterplot?

4.1. Table 4.15 shows the batting average and number of runs per game for all of the National League teams in the 1999 season.

TABLE 4.15
Batting average and runs scored per game for 1999 National League teams.

Team	AVG	Runs per Game
COL	0.288	5.59
NY	0.279	5.23
AZ	0.277	5.60
PHI	0.275	5.19
MIL	0.273	5.06
CIN	0.272	5.30
SF	0.271	5.38
HOU	0.267	5.08
ATL	0.266	5.18
LA	0.266	4.89
MON	0.265	4.43
FLA	0.263	4.26
STL	0.262	5.02
PIT	0.259	4.81
CHI	0.257	4.61
SD	0.252	4.38

(a) On the grid below, construct a scatterplot of Runs per Game (vertical axis) against Batting Average (horizontal).

(b) What does the scatterplot say about the relationship between Runs per Game and Batting Average? If you know that a team has a high batting average, what does that say about the number of runs it scores?

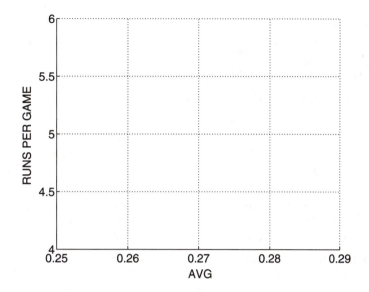

4.2. (Exercise 4.1 continued)   A least-squares fit to the (Runs per Game, Batting Average) data for the 1999 NL teams gives the relationship.

$$\text{runs} = -4.69 + 36.1 \text{ AVG}.$$

(a) Suppose that a team has a .250 batting average. How many runs do you predict the team will score in a game?

(b) Suppose that a team has a .260 batting average. Predict the number of runs the team will score.

(c) Suppose a team hits for a 10-point higher batting average (say from .250 to .260). How many additional runs per game could we have predicted the team to score?

(d) St. Louis in 1999 had a .262 batting average and scored an average of 5.02 runs per game. For St. Louis, find the predicted runs scored and the residual. Can you explain why the residual is positive in this case?

4.3. Table 4.16 gives a number of pitching statistics for the 1999 National League teams. The abbreviations used in the table are explained following.

SOR = number of pitching strikeouts per game

BOR = number of pitching walks per game

ERA = earned run average

PCT = winning percentage

HG = number of hits allowed per game

HRG = number of home runs allowed per game

RG = number of runs allowed per game

TABLE 4.16

Team pitching statistics for 1999 National League teams.

Team	SOR	BOR	ERA	PCT	HG	HRG	RG
COL	6.37	3.13	6.03	0.444	10.49	1.46	6.34
NY	7.19	4.39	4.27	0.595	8.41	0.98	4.36
AZ	7.39	3.62	3.77	0.617	8.56	1.08	4.17
PHI	6.35	3.89	4.93	0.475	9.22	1.30	5.22
MIL	6.13	4.08	5.08	0.46	10.04	1.32	5.50
CIN	6.63	3.49	3.99	0.589	8.03	1.16	4.36
SF	6.64	4.29	4.71	0.531	9.17	1.19	5.12
HOU	7.43	4.49	3.84	0.599	9.16	0.79	4.16
ATL	7.38	3.75	3.65	0.636	8.62	0.87	4.08
LA	6.64	3.66	4.45	0.475	8.87	1.18	4.85
MON	6.43	2.70	4.69	0.420	9.29	0.93	5.26
FLA	5.82	2.95	4.90	0.395	9.62	1.05	5.25
STL	6.36	3.8	4.76	0.466	9.43	1.00	5.20
PIT	6.72	3.55	4.35	0.484	8.96	0.99	4.85
CHI	6.04	3.52	5.27	0.414	9.99	1.36	5.67
SD	6.65	3.89	4.47	0.457	8.97	1.19	4.82

(a) Circle the variables below that you believe have a *positive* association with runs allowed per game (RG).

SOR   BOR   ERA   PCT   HG   HRG

(b) Circle the variables below that you believe have a *negative* association with runs allowed per game (RG).

SOR   BOR   ERA   PCT   HG   HRG

(c) Do you think a team's pitched strikeouts per game (SOR) is related to a team's pitched walks per game (BOR)? Explain what type of relationship you would expect to find and why.

(d) What variable among the above do you think has the *strongest* relationship with a team's winning percentage (PCT)?

4.4. (1999 NL team pitching data)   Figure 4.15 displays scatterplots of runs allowed per game (RG) and each of the variables SOR, BOR, ERA, PCT, HG, HRG.

(a) From looking at the scatterplots, list two variables that have a negative association with runs allowed per game.

(b) List two variables that have a positive association with runs allowed per game.

(c) List the variables in order in terms of their association with runs allowed, from most negatively associated to most positively associated.

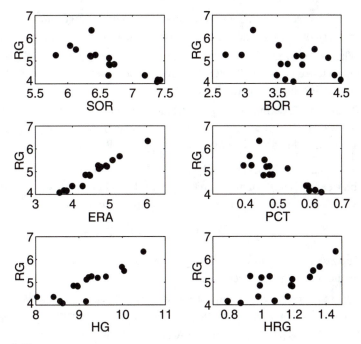

FIGURE 4.15
Scatterplots of runs allowed and different pitching statistics for 1999 National League teams.

4.5. (1999 NL team pitching data)   Table 4.17 gives the correlations between all of the variables in Exercise 4.6 for the 1999 NL team pitching data. To help interpret the table, note that the correlation between the walk rate (BOR) and the strikeout rate (SOR) is +.489. This means that there is a positive association between the strikeout and walk rates—teams that struck out a lot of batters tended also to walk a lot of batters, and likewise teams with small strikeout rates tended also to have small walk rates.

TABLE 4.17
Correlation matrix for a number of pitching variables for 1999 National League teams.

	SOR	BOR	ERA	PCT	HG	HRG
BOR	0.489					
ERA	−0.784	−0.347				
PCT	0.899	0.543	−0.790			
HG	−0.650	−0.264	0.870	−0.720		
HRG	−0.597	−0.144	0.756	−0.481	0.535	
RG	−0.812	−0.425	0.986	−0.832	0.885	0.727

(a) From looking at the correlation matrix, which variable has the strongest relationship with the average runs per game allowed (RG)? Why do you think the correlation value is so close to one?

(b) Which variable has the weakest association with average runs per game? Does this variable have a positive or negative association with RG?

(c) What is the correlation between home runs allowed per game (HRG) and strike-outs pitched per game (SOR)? Explain in words what this correlation says about the relationship between HRG and SOR.

(d) By the correlation matrix, the correlation between home runs allowed per game and walks allowed per game is -0.144. Explain in words what this means about the relationship between home runs and walks allowed.

4.6. (1999 NL team pitching data)   We saw earlier that there was a negative relationship between a team's winning percentage (PCT) and the team's earned run average (ERA). A least-squares line through the data has the equation

$$PCT = 0.967 - 0.101 \text{ ERA.}$$

(a) Suppose a team allows, on average, four earned runs per game. Use the line to predict its winning percentage.

(b) Suppose a team's ERA jumps from 4.00 to 5.00. Using the least-squares line, do you predict its winning percentage would increase or decrease? By how much?

(c) Colorado in 1999 had a 6.03 ERA and a .444 winning percentage. Find Colorado's predicted PCT and the residual. Can you explain why the residual is large in this case?

4.7. Two hundred teams were selected randomly from baseball history. For each team, the season and the number of triples hit per game were recorded. Figure 4.16 plots the triples per game (vertical) against the season (horizontal).

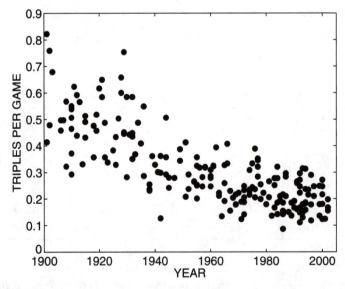

FIGURE 4.16
Scatterplot of triples per game against season for 200 randomly selected baseball teams.

(a) Describe the basic pattern in the graph. Has the number of triples per game increased or decreased over time?

(b) Would it be appropriate to find a best line to these data? Why or why not?

(c) Can you offer any explanation for the pattern in this graph? In other words, why has the frequency of triples changed over the years?

4.8. Two hundred teams were selected randomly from baseball history. For each team, the season and the number of stolen bases (SB) per game were recorded. In Figure 4.17, we plot the SB per game (vertical) against the season (horizontal).

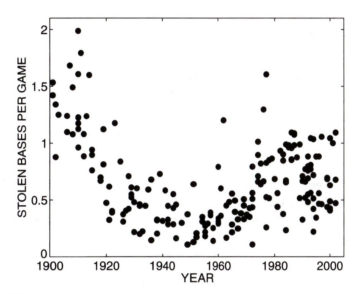

FIGURE 4.17
Scatterplot of stolen bases per game against season for 200 randomly selected baseball teams.

(a) Describe the basic pattern in the graph. Is the number of SBs per game increasing or decreasing over time?

(b) Is there a "straight-line" relationship between SB per game and year?

(c) What does this graph say about the importance of stolen bases in baseball today compared to the past?

4.9. For all pitchers in MLB history, we collected the year of birth and the throwing hand (left or right). For each birth year, the fraction of left-handed throwers is computed and Figure 4.18 displays a time series plot of the fraction as a function of the birth year.

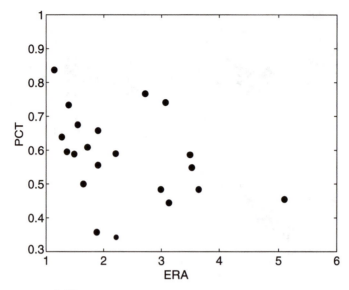

FIGURE 4.20
Scatterplot of season ERA and winning percentage for Walter Johnson.

(c) Based on this graph, do you think that PCT is a good measure of a pitcher's performance? Why or why not?

(d) A famous pitcher who played primarily for Texas teams was recently inducted into the Hall of Fame but had a winning percentage that was close to 50%. Name this pitcher. Explain why he was elected in spite of his low winning percentage.

4.12. Over the last hundred years, there has been a substantial growth in the United States population, and a corresponding large growth in the number of players and fans of Major League baseball. Table 4.18 gives the U.S. population and total MLB baseball

TABLE 4.18
United States population, Major League baseball attendance, and ratio for the years 1900 through 1990.

Year	Population	Baseball Attendance	Attendance/ Population
1900	76,212,168	1,829,490	0.024
1910	92,228,496	6,206,447	0.067
1920	106,021,537	9,120,875	0.086
1930	123,202,624	10,132,262	0.082
1940	132,164,569	9,823,484	0.074
1950	151,325,798	17,462,977	0.115
1960	179,323,175	19,911,489	0.111
1970	203,302,031	28,747,333	0.141
1980	226,542,199	43,014,136	0.190
1990	248,709,873	54,823,768	0.220

attendance for the years 1900, 1910, ..., 1990. The third column divides the baseball attendance by the population size. We see that in 1990, baseball attendance was equal to 22 percent of the American population.

Figure 4.21 plots the attendance/population ratio as a function of year.

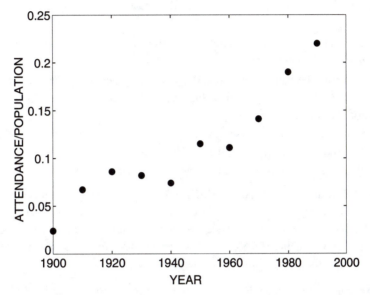

FIGURE 4.21
Time series plot of the ratio of Major League baseball attendance to population.

(a) Describe, using a few sentences, how the attendance/population ratio has changed over the years. Would it be accurate to say that the ratio has consistently increased from 1900 to 1990?

(b) Would it be reasonable to fit a least-squares line to these data? Why or why not?

(c) What would explain the dip in the ratio at the year 1940? (*Hint:* What was happening in the world scene at this time?)

(d) If one fits a least-squares line only for the 1960–1990 data, one obtains the equation

$$\text{Ratio} = -7.2605 + 0.0038 \, (\text{Year}).$$

Use this equation to predict the attendance/population ratio for the year 2000.

(e) Would you expect the same straight-line relationship between Ratio and Year for the next 50 years? Why or why not?

4.13. Suppose you are interested in comparing the batting abilities of Wade Boggs and Tony Gwynn for the 1987 baseball season; the batting statistics for both players are given in Table 4.19.

TABLE 4.19

Batting statistics of Wade Boggs and Tony Gwynn for the 1987 season.

	AB	R	H	2B	3B	HR	BB	1B	TB	AVG	SLG	OBP	OPS	RC
Wade Boggs	551	108	200	40	6	24	105							
Tony Gwynn	589	119	218	36	13	7	82							

(a) Compute the number of singles (1B) and total bases (TB) for each player and place the values in the table.

(b) Compute the batting average (AVG), slugging percentage (SLG), on-base percentage (OBP), on-base plus slugging (OPS), and runs created (RC) measures for each player.

(c) Compare the two players with respect to ability to get on base, slugging ability, and total offensive contribution for this season.

(d) Which player would you rather have on your team? Why?

4.14. Find the batting statistics (AB, R, H, 2B, 3B, HR, BB) for the most recent year for two good hitters, one from the National League and one from the American League. Find the batting average (AVG), slugging percentage (SLG), on-base percentage (OBP), on-base plus slugging (OPS), and runs created (RC) measures for each player. Put your stats in the table below. Which player had a better offensive year? Why?

NAME	AB	R	H	2B	3B	HR	BB	1B	TB	AVG	SLG	OBP	OPS	RC

4.15. (Least-squares criterion)   Suppose you are interested in a typical number of home runs hit by a National League team in 1999, and that you guess that the typical number is 150 home runs. We can evaluate the goodness of the guess 150 by means of the Root Mean Squared Error (RMSE) criterion. Table 4.20 summarizes the calculations of RMSE using the 1999 National League team home run data. (The Residual column contains the difference between the HR value and the guess, and the Residual^2 column contains the square of the residual.) Fill in the missing cells in the table and compute the RMSE.

4.16. (Exercise 4.15 continued)   In the previous exercise, we made a guess that a typical number of home runs hit by an NL team in 1999 was 150. In fact, the best guess at this typical number of home runs is the mean, which is found by adding up all of the home run numbers and dividing by the number of teams (16). Here the mean is

$$\bar{x} = \frac{223 + 181 + \cdots + 189 + 153}{16} = 181.$$

(a) Compute the RMSE of the mean by completing Table 4.21.

(b) Compare the RMSE of the mean with the RMSE of the guess 150 that you found in Exercise 4.16. Which is a better estimate at a typical home run total?

TABLE 4.20

Summary of RMSE calculations for the 1999 National League team home run data using the guess 150.

Team	HR	Gyess	Residual	$Residual^2$
COL	223	150		
NY	181	150	31	961
AZ	216	150	66	4356
PHI	161	150		
MIL	165	150	15	225
CIN	209	150	59	3481
SF	188	150	38	1444
HOU	168	150	18	324
ATL	197	150		
LA	187	150	37	1369
MON	163	150	13	169
FLA	128	150	−22	484
STL	194	150		
PIT	171	150	21	441
CHI	189	150	39	1521
SD	153	150	3	9
SUM	xxxxxxx	xxxxxxx	xxxxxxx	24379

TABLE 4.21

Summary of RMSE calculations for the 1999 National League team home run data using the mean estimate 181.

Team	HR	Guess	Residual	$Residual^2$
COL	223	181	42	1764
NY	181	181	0	0
AZ	216	181	35	1225
PHI	161	181		
MIL	165	181	−16	256
CIN	209	181		
SF	188	181	7	49
HOU	168	181	−13	169
ATL	197	181	16	256
LA	187	181		
MON	163	181	−18	324
FLA	128	181	−53	2809
STL	194	181		
PIT	171	181	−10	100
CHI	189	181	8	64
SD	153	181	−28	784
SUM	xxxxxxx	xxxxxxx	xxxxxxxxx	9189

TABLE 4.22

RMSE calculations using AVG to predict runs per game.

Team	AVG	Runs Per Game	Predicted	Residual	$Residual^2$
COL	0.288	5.59	5.71	−0.12	0.0144
NY	0.279	5.23	5.38	−0.15	0.0225
AZ	0.277	5.60	5.31		
PHI	0.275	5.19	5.24		
MIL	0.273	5.06	5.17		
CIN	0.272	5.30	5.13		
SF	0.271	5.38	5.09		
HOU	0.267	5.08	4.95	0.13	0.0169
ATL	0.266	5.18	4.91	0.27	0.0729
LA	0.266	4.89	4.91	−0.02	0.0004
MON	0.265	4.43	4.88	−0.45	0.2025
FLA	0.263	4.26	4.80	−0.54	0.2916
STL	0.262	5.02	4.77	0.25	0.0625
PIT	0.259	4.81	4.66	0.15	0.0225
CHI	0.257	4.61	4.59	0.02	0.0004
SD	0.252	4.38	4.41	−0.03	0.0009
SUM					

4.17. (Exercise 4.1 continued) The AVG and Runs Per Game columns of Table 4.22 give the batting averages and runs per game for the 1999 NL teams. The Predicted column gives the predicted runs per game using the least-squares fit Runs = −4.69 + 36.1 AVG. The RESIDUAL column gives the residual for each team, and the Residual^2 column contains the squared value of the residual.

(a) Fill in the missing cells in the Residual and Residual^2 columns.

(b) Compute the Root Mean Square (RMSE) value. This is a measure of the goodness of using AVG as a predictor of runs scored per game.

4.18. (Exercise 4.1 continued) Suppose that the slugging percentage (SLG) is used, instead of AVG, to predict runs per game. The least-squares fit is given by

$$\text{Runs per Game} = -2.58 + 17.7 \text{ SLG}.$$

Table 4.23 contains the predicted runs per game, the residuals, and the residuals squared using SLG as a predictor.

(a) Fill in the missing cells in the Residual and Residual^2 columns.

(b) Compute the RMSE value for the SLG fit. Compare this value with the RMSE value from the AVG fit that you computed in Exercise 4.17.

(c) Construct parallel boxplots of the residuals from the SLG fit and the residual from the AVG fit (Exercise 4.17). Looking at the boxplot display, which is the better predictor—AVG or SLG? Why?

TABLE 4.23
RMSE calculations using SLG to predict runs per game.

Team	Runs Per Game	SLG	Predicted	Residual	$Residual^2$
COL	5.59	0.472	5.77	−0.18	0.0324
NY	5.23	0.434	5.10	0.13	0.0169
AZ	5.60	0.459	5.54	0.06	0.0036
PHI	5.19	0.431	5.05	0.14	0.0196
MIL	5.06	0.426	4.96	0.10	0.0100
CIN	5.30	0.451	5.40		
SF	5.38	0.434	5.10		
HOU	5.08	0.420	4.85		
ATL	5.18	0.436	5.14		
LA	4.89	0.420	4.85	0.04	0.0016
MON	4.43	0.427	4.98	−0.55	0.3025
FLA	4.26	0.395	4.41	−0.15	0.0225
STL	5.02	0.426	4.96	0.06	0.0036
PIT	4.81	0.419	4.84	−0.03	0.0009
CHI	4.61	0.420	4.85	−0.24	0.0576
SD	4.38	0.393	4.38	0.00	0.0000
SUM					

TABLE 4.24
RMSE calculations using OPS to predict runs per game.

Team	Runs Per Game	OBP	SLG	OPS	Predicted	Residual	$Residual^2$
COL	5.59	0.348	0.472	0.820	5.68	−0.09	0.0081
NY	5.23	0.363	0.434	0.797	5.35	−0.12	0.0144
AZ	5.60	0.347	0.459	0.806	5.48	0.12	0.0144
PHI	5.19	0.351	0.431	0.782	5.13	0.06	0.0036
MIL	5.06	0.353	0.426	0.779	5.09	−0.03	0.0009
CIN	5.30	0.341	0.451	0.792	5.28	0.02	0.0004
SF	5.38	0.356	0.434				
HOU	5.08	0.355	0.420				
ATL	5.18	0.341	0.436				
LA	4.89	0.339	0.420	0.759	4.80	0.09	0.0081
MON	4.43	0.323	0.427	0.750	4.68	−0.25	0.0625
FLA	4.26	0.325	0.395	0.720	4.25	0.01	0.0001
STL	5.02	0.338	0.426	0.764	4.88	0.14	0.0196
PIT	4.81	0.334	0.419	0.753	4.72	0.09	0.0081
CHI	4.61	0.329	0.420	0.749	4.66	−0.05	0.0025
SD	4.38	0.332	0.393	0.725	4.32	0.06	0.0036
							0.1801

4.19. (Exercise 4.1 continued)   Suppose that we estimate the runs per game using the derived statistic OPS = OBP + SLG as a predictor. The least-squares fit is given by

$$\text{runs} = -6.05 + 14.3 * \text{OPS}.$$

(a) Suppose a team has an OBP = .360 and SLG = .700. What is the value of the team's OPS?

(b) For each team in the 1999 NL, Table 4.24 computes the OPS statistic, the predicted runs per game (using the least-squares formula), the residual and the residual squared.

(c) Fill in the blank cells in the table.

(d) Compute the RMSE. Compare the RMSE value with the RMSE values using the predictors AVG and SLG (Exercises 4.3 and 4.4). Which is the best predictor among AVG, SLG, and OPS? Why?

Further Reading

Devore and Peck (2000) and Moore and McCabe (1998) describe basic descriptive tools for understanding their relationship between two measurement variables. Albert (1998), Bennett (1998), Thorn and Palmer (1985), and Albert and Bennett (2001), Chapters 6, 7, 8, describe a number of ways of measuring offensive performance. The mean squared error criterion for evaluating a particular batting measure is used in Chapter 6 of Albert and Bennett (2001). The Pythagorean Formula for relating a team's win/losses with the number of runs scored/allowed is described in James (1982, 2001). A nice discussion of the regression-to-the-mean effect is given in Berry (1996).

5

Introduction to Probability Using Tabletop Games

What's On-Deck?

In this chapter we introduce basic concepts of probability by using tabletop games. To begin, we introduce the relative frequency notion of probability by using Barry Bonds's hitting log for the 2001 season. During a plate appearance, Bonds can either hit a home run or not, and we assume that the chance that he hits a home run is p. We learn about the value of p by observing his hitting performance over a number of games and we can estimate his home run probability by calculating the relative frequency of home runs. In Case Study 5-2, we explore a tabletop dice game, *Big League Baseball*, and find probabilities of different events, say home runs, singles, and outs, by finding probabilities of the sum of two rolls of the dice. The game *All Star Baseball* is a more elaborate game in that the batting performance of each hitter is modeled using a separate spinner. In Case Study 5-3, we show how career statistics for a player can be used to compute probabilities that the player gets a single, double, etc., and these probabilities are used to compute areas of the random spinner. We conclude our discussion by describing a more sophisticated game, *Strat-O-Matic Baseball*, that uses four dice and models the abilities of each pitcher and each hitter by a separate card. We consider a classic matchup in this game—Mark McGwire against Greg Maddux—and show how one can compute the probability of McGwire hitting a home run using the theorem of total probabilities.

Case Study 5-1: What Is Barry Bonds's Home Run Probability?

Topics Covered: Relative frequency interpretation of probability, law of large numbers
In this case study, we begin our discussion on probability.

What is a probability? First, we recognize that life is full of uncertain events. For example,

- Who will win the next World Series?

- Will you retire before the age of 60?

- Will a Major League player ever break Joe DiMaggio's record of hitting in 56 consecutive games?

A probability is a way of measuring the uncertainty that we see. We can define a probability scale from 0 to 1 and any event can be assigned a number in this range.

- We assign a probability of 0 to an event that we are certain will not occur. For example, the probability that the Phillies will meet the Mets in the World Series is zero since the World Series matches the winners of the National League and American League playoffs and the Phillies and Mets both play in the National League.

- On the other hand, we assign a probability of 1 to an event that we are sure will occur, such as "a Major League Baseball game will finish in ten hours."

- What if we assign a probability of .5? Suppose we toss a coin. We give the event "heads" a probability of .5 if we think that the events "heads" and "tails," or "not heads," have the same chance of occurring.

One way of thinking about probabilities is the following *relative frequency* interpretation. An *experiment* is a process where the outcome (or result) is unknown. We let an *event* be a collection of outcomes, and we are interested in computing the probability of the event.

Say that we can repeat the experiment many times, say N times, under similar conditions. For example, if the experiment is tossing a coin, then suppose that we can toss the coin repeatedly under similar conditions. Then the Prob(event) is approximately the relative frequency of the event. That is,

$$\text{Prob(event)} \approx \frac{\text{number of times event occurs}}{N}.$$

To illustrate this interpretation, suppose we are interested in estimating the probability that Barry Bonds in 2001 would hit a home run in a single plate appearance. We assume that Barry comes to bat multiple times during the season under the same conditions. (Note that this is a questionable assumption, but it simplifies our discussion.) For our purposes, there are two results of this plate appearance—he either hits a home run or he doesn't, and the chance that he hits a home run in this particular season is measured by a probability p_{HR}.

We don't know the value of Bonds's home run probability p_{HR}, but we can learn about it by watching Bonds perform during the 2001 baseball season. Table 5.1 gives the number of plate appearances (PA) and the number of home runs (HR) Barry hit for each of the 153 games he played in 2001.

In the first game, Barry came to bat five times and had a single home run. At that point, the current estimate of Bonds's home run probability is

$$\hat{p}_{HR} = \tfrac{1}{5} = .200.$$

TABLE 5.1
Plate appearances and home runs for Barry Bonds for each game played during the 2001 baseball season.

Game	PA	HR	Est.	Game	PA	HR	Est.	Game	PA	HR	Est.
1	5	1	1/5 = .200	52	3	0		103	3	1	
2	5	0		53	4	1		104	5	0	
3	4	0		54	5	1		105	6	1	
4	4	0		55	5	1		106	5	0	
5	4	0		56	4	0		107	5	1	
6	5	0		57	5	0		108	5	1	
7	5	0		58	4	0		109	5	0	
8	4	1		59	4	1		110	5	1	
9	5	1		60	4	0		111	4	0	
10	5	1	4/46 = .087	61	5	1		112	4	2	
11	4	1		62	4	2		113	4	0	
12	4	1		63	3	0		114	4	1	
13	4	1		64	4	0		115	4	0	
14	3	0		65	7	1		116	5	0	
15	4	1		66	5	1		117	4	0	
16	1	0		67	5	0		118	1	1	
17	4	0		68	5	1		119	4	0	
18	4	1		69	5	0		120	5	0	
19	5	0		70	4	0		121	4	0	
20	5	1		71	5	0		122	5	1	
21	3	0		72	5	0		123	4	0	
22	4	1		73	4	0		124	4	0	
23	5	0		74	4	0		125	5	0	
24	5	1		75	4	0		126	4	1	
25	5	1		76	5	0		127	1	0	
26	4	1		77	5	0		128	4	0	
27	5	0		78	5	0		129	4	1	
28	1	0		79	5	0		130	4	1	
29	4	0		80	3	0		131	3	0	
30	4	0		81	7	0		132	5	1	
31	3	0		82	5	1		133	6	0	
32	5	1		83	5	0		134	5	0	
33	4	0		84	4	0		135	6	3	
34	4	0		85	5	0		136	4	0	
35	5	0		86	4	0		137	4	0	

(*continued*)

TABLE 5.1
Continued.

Game	PA	HR	Est.	Game	PA	HR	Est.	Game	PA	HR	Est.
36	4	0		87	4	0		138	5	1	
37	4	1		88	3	2		139	4	0	
38	4	1		89	5	0		140	4	0	
39	5	3		90	4	0		141	5	2	
40	4	2		91	4	0		142	5	1	
41	4	1		92	3	0		143	5	0	
42	5	1		93	4	0		144	5	0	
43	4	0		94	4	0		145	5	1	
44	4	1		95	1	0		146	4	1	
45	4	0		96	5	2		147	4	0	
46	4	1		97	5	1		148	5	0	
47	5	0		98	5	0		149	5	0	
48	8	0		99	4	0		150	5	1	
49	4	2		100	6	0		151	5	2	
50	5	1		101	4	1		152	1	0	
51	5	0		102	4	0		153	4	1	$73/664 =$ 0.1099

(We put this estimate in the table.) Certainly, this is not a great estimate of Bonds's home run probability since it is based on only five plate appearances. Next, suppose it is the middle of April and we've now watched Barry play ten games. In games 1–10, he has 46 PA and 4 HR—our new estimate of Bonds's home run probability is

$$\hat{p}_{HR} = \tfrac{4}{46} = .087.$$

This is likely a better estimate of Bonds's ability since it is based on a greater number of plate appearances. Suppose that we compute Bonds's 2001 home run rate after each of his 153 games. Figure 5.1 plots the home run rate (our probability estimate) against the game number.

Note that for early game numbers, the probability estimate shows a lot of fluctuation. But after about game 80, the home run rate appears to settle down to about the value 0.11. This illustrates the relative frequency of probability. As Bonds gets more and more plate appearances, the relative frequency of his home runs settles down and approaches his probability of a home run p_{HR} for 2001. After 153 games, he hit 73 home runs in 664 plate appearances for an estimate of $73/664 = .1099$. This value is a good estimate of Bonds's home run ability for the 2001 season.

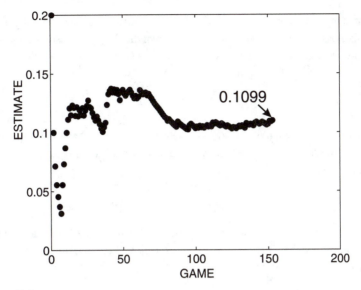

FIGURE 5.1
Plot of home run rate against game number for Barry Bonds for the 2001 baseball season.

Case Study 5-2: *Big League Baseball*

Topics Covered: Sample space, experiments with equally likely outcomes, experiment of rolling two dice, finding probabilities of events   Assigning probabilities is generally hard to do if we are not able to repeat the experiment many times. But probabilities can be assigned for simple experiments, such as those involving dice and cards, and we talk about several of these experiments here. We describe a simple 1960s tabletop game, *Big League Baseball*, based on dice rolls.

Suppose we roll a fair die. We first think of the possible outcomes of doing this. There are six possibilities:

| Roll | 1 | 2 | 3 | 4 | 5 | 6 |

We call the collection of all possible outcomes the *sample space*.

To assign probabilities, we assume that each roll outcome has the same probability. (In other words, we assume that the outcomes are *equally likely*.) We want to assign a positive number to each outcome such that the sum of all the probabilities assigned will be equal to 1, since we are certain there will be one of the six outcomes in each roll. It should be clear that we should assign a probability of $\frac{1}{6}$ to each outcome:

Roll	1	2	3	4	5	6
Probability	$\frac{1}{6}$	$\frac{1}{6}$	$\frac{1}{6}$	$\frac{1}{6}$	$\frac{1}{6}$	$\frac{1}{6}$

Next suppose we roll two dice. It will be convenient to distinguish the dice, so we will take one die to be brown and one to be orange. (Brown and orange are the colors of the author's university.)

1. How many outcomes are there in this experiment? We know from above that there are six possibilities for the result on the brown die. For each result on the brown die, say brown $= 2$, there are six possible outcomes for the orange die. So the number of possible outcomes for two dice is

$$6 \times 6 = 36$$

2. If we assume that each of the 36 possible rolls of two dice have the same probability, then (by the same logic as before) we assign a probability of $\frac{1}{36}$ to each outcome. So, for example,

$$\text{Prob(5 on brown and 2 on orange)} = \frac{1}{36}.$$

Big League Baseball was a dice tabletop baseball game made by Sycamore Games of Lima, Ohio in the 1960s. This game is based on rolling 1 red die and 2 white dice. First, roll the red die to get the pitch:

- If 1 or 6 is rolled, a fair ball is hit.

- If 2 or 3 is rolled, a ball is pitched

- If 4 or 5 is rolled, a strike is pitched.

Let's find the probabilities of some outcomes:

1. The probability of pitching a ball, Prob(ball), is the same as the probability of rolling a 2 or 3. We find the probability of a set of outcomes by *adding* the probabilities of the outcomes. So

$$\text{Prob(ball)} = \text{Prob(2 or 3)} = \text{Prob(2)} + \text{Prob(3)} = \frac{1}{6} + \frac{1}{6} = \frac{2}{6}.$$

2. The probability of pitching a ball or a strike is found the same way:

$$\text{Prob(ball or strike)} = \text{Prob(2, 3, 4, 5)} = \text{Prob(2)} + \text{Prob(3)} + \text{Prob(4)} + \text{Prob(5)} = \frac{4}{6}.$$

3. What is the probability that a strike is *not* thrown? We note that a strike is not thrown if a ball is pitched or a fair ball is hit, so

$$\text{Prob(no strike)} = \text{Prob(ball or fair ball)} = \text{Prob(1, 6, 2, 3)} = \frac{4}{6}.$$

In *Big League Baseball*, if a ball is in play, then two white dice are rolled. Table 5.2 shows the outcomes for each possibility of rolling two dice.

Since each cell in the table is assigned a probability of $\frac{1}{36}$, we can compute several probabilities of interest:

   (a) Prob(Home run) $= \frac{1}{36}$. (There is just one way to roll a home run.)

   (b) Prob(Single). We see from the table that there are 7 ways of getting a single—each outcome has probability $\frac{1}{36}$, so Prob(Single) $= \frac{7}{36}$.

   (c) Prob(Hit) $= \frac{10}{36}$. (From the table, we see 10 ways of getting a hit.)

TABLE 5.2

Outcomes of the rolls of two dice.

		2nd die					
		1	2	3	4	5	6
	1	Single	Out	Out	Out	Out	Error
	2	Out	Double	Single	Out	Single	Out
1st	3	Out	Single	Triple	Out	Out	Out
die	4	Out	Out	Out	Out	Out	Out
	5	Out	Single	Out	Out	Out	Single
	6	Error	Out	Out	Out	Single	Home run

(d) Prob(Out) $= \frac{24}{36}$. (It is easiest to note that there are 12 ways of getting on base by a hit or error, and therefore $36 - 12 = 24$ ways of getting an out.)

Is *Big League Baseball* a realistic baseball game? In other words, does this game provide a good representation of real baseball? Of course not—it would be insulting to the game to think that we could simulate real baseball by using only three dice. This game assumes many unrealistic things, including

- that all batters have the same ability,

- that all pitchers have the same ability,

- that it is equally likely to add a strike or a ball to the pitch count. (This is not realistic—in real baseball, it is more common to pitch a strike than a ball.)

Although this game is a bit unrealistic, it is a nice first attempt to simulate a baseball game. The results of a game are partly due to chance variation, and this game introduces chance variation by the use of dice. This game provides a useful comparison to the more sophisticated baseball games that are described in the next two case studies.

Case Study 5-3: *All Star Baseball*

Topics Covered: Spinner as a randomization device, probabilities represented by areas of the spinner, multinomial experiment   We next consider a more sophisticated baseball game, *All Star Baseball*. This game was popular in the 1960s and 1970s and was played by the author when he was young.

This game is based on using a spinner. Each hitter is represented by a circular spinner where different areas of the spinner correspond to different play outcomes. An *All Star Baseball* spinner for the great baseball player Babe Ruth is displayed in Figure 5.2. The areas of the spinner correspond to the probabilities of the different outcomes. This randomization device has more flexibility than dice, since we can represent a greater range of probabilities in the outcomes.

We illustrate constructing a spinner for one of my favorite players, Mike Schmidt. By the way, it is doubtful that one spinner for an entire career is realistic, but we need to start

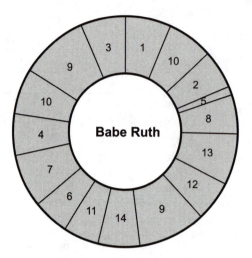

1: HR   7,13: 1B   11: 2B   5: 3B   9: BB
10: K   2,6,12: Ground Ball   3,4,8,14: Fly Ball

FIGURE 5.2

All Star Baseball spinner for Babe Ruth.

somewhere. We start with a very simple model—one spinner for an entire career, regardless of age, ballpark, opposing pitcher, and defense. In later chapters, we will examine the benefits and drawbacks of using more complicated models.

We make this spinner in two steps. First, we calculate approximate probabilities for the different outcomes (1B, 2B, 3B, HR, BB, Out) that can happen when Schmidt comes to bat. Then we shade regions on the spinner where the areas of the regions correspond to the probabilities. We start with Schmidt's career statistics shown below.

Mike Schmidt

G	AB	R	H	2B	3B	HR	SO	BB	AVG	OBP	SLG
2404	8352	1506	2234	408	59	548	1883	1507	.267	.380	.527

We want to classify all plate appearances (PAs) into the outcomes 1B, 2B, 3B, HR, BB, and Outs. We ignore events like HBP and Sacrifices, since they are relatively insignificant compared to the other outcomes. We put the counts we know from the career data in Table 5.3.

We complete Table 5.3 in Table 5.4, using the computations below. We get PAs by adding at-bats (AB) and walks (BB).

$$PA = AB + BB = 8352 + 1507 = 9859.$$

We get singles (1B) by adding the doubles (2B), triples (3B), and home runs (HR), and subtracting the total from hits (H).

TABLE 5.3
Counts of career offensive statistics for Mike Schmidt.

Play	Count
PA	
AB	8352
H	2234
1B	
2B	408
3B	59
HR	548
BB	1507
Outs	

TABLE 5.4
Counts of career offensive statistics for Mike Schmidt with PA, singles, and outs included.

Play	Count
PA	9859
AB	8352
H	2234
1B	1219
2B	408
3B	59
HR	548
BB	1507
Outs	6118

$$1B = H - (2B + 3B + HR) = 2234 - (408 + 59 + 548) = 1219.$$

We get outs by subtracting hits (H) from at-bats (AB).

$$Outs = AB - H = 8352 - 2234 = 6118.$$

We next change these counts to approximate probabilities in Table 5.5 by dividing each count (1B, 2B, 3B, HR, BB, OUTS) by the number of PAs. We check if we did this right by seeing if the sum of probabilities is equal to 1.

To make our spinner, we start with a blank circle and shade areas of the regions of single (1B), double (2B), etc., that correspond to these probabilities. To make the construction process easier, one can divide the circular region into 36 equal regions. We convert the probabilities into number of regions by multiplying the probability by 36 and rounding the result to the nearest whole number:

$$Number\ of\ Regions = round\ (36 \times probability).$$

TABLE 5.5
Computation of event probabilities from Mike Schmidt's career offensive statistics.

Event	Count	Proportion
PA	9859	xxxxx
AB	8352	xxxxx
H	2234	xxxxx
1B	1219	0.124
2B	408	0.041
3B	59	0.006
HR	548	0.056
BB	1507	0.153
Outs	6118	0.621
		1

To illustrate, we multiply the probability of a single by 36 to get

$$\text{Number of regions} = 36 \times 0.124 = 4.46.$$

We round this to the nearest integer, getting four regions. If we do this calculation for all events, we get the region numbers in Table 5.6.

TABLE 5.6
Computation of spinner region numbers from Mike Schmidt's career offensive statistics.

Event	Count	Proportion	Region
PA	9859	xxxx	xxxx
AB	8352	xxxx	xxxx
H	2234	xxxx	xxxx
1B	1219	0.124	4
2B	408	0.041	1
3B	59	0.006	0
HR	548	0.056	2
BB	1507	0.153	6
Outs	6118	0.621	22
		1	35

This didn't quite work, since the total number of regions is 35, not 36. So we make a small adjustment—I changed the number of regions of 1B from 4 to 5—to make the sum of regions add up to 36. (I adjusted the number of regions of 1B instead of HR, say, since this adjustment has a modest change in the single probability.)

We completed the calculations for our spinner and now can begin the construction process. We take a blank spinner, shown in Figure 5.3, and color-code the events according to the work in Table 5.6. So we color five regions black (corresponding to single), one region

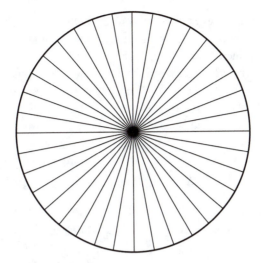

FIGURE 5.3
Blank spinner divided into 36 equal-sized regions.

purple (corresponding to double), two regions light purple, six regions green (walk) and 22 regions white (out). When we are done we get the spinner in Figure 5.4. We will be using a spinner as a basic probability model in our study of inference in Chapter 7.

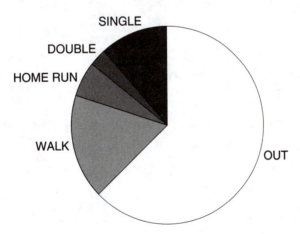

FIGURE 5.4
Spinner using Mike Schmidt's career hitting statistics.

Case Study 5-4: *Strat-O-Matic Baseball*

Topics Covered: Probabilities of the sum of two dice, theorem of total probabilities, conditional probability   We conclude our discussion of tabletop games by briefly describing my favorite game, *Strat-O-Matic Baseball*. It's a game played with three dice (like *Big League Baseball*). The game is much more realistic than the two previous games discussed, since it models the different abilities of pitchers as well as hitters. The *All-Star Baseball* game described in Case Study 5-3 doesn't take into account the abilities of pitchers.

We introduce this game by considering a classic matchup between Mark McGwire and Greg Maddux. (McGwire was arguably the greatest power hitter and Maddux the greatest control pitcher in recent baseball history.) Each player is represented by a card—the Mark McGwire and Greg Maddux cards are shown in Figure 5.5. We first roll a single white die. If the result is 1, 2, 3, we look at McGwire's card; otherwise we look at Maddux's card. Then we roll two red dice and observe the sum. The play is determined by reading the line (corresponding to the sum) on the pitcher's or hitter's card. Sometimes the outcome is not determined by the roll of the white and red dice and a twenty-sided die must be rolled to determine the play result.

MARK McGWIRE 1st base-3 stealing-E running 1-8

ST. LOUIS

1	2	3
2-lineout (ss) into as many outs as possible	2-flyball (cf)B	2-WALK
3-strikeout	3-WALK	3-WALK
4-flyball (lf)B	4-HOMERUN	4-strikeout
5-WALK	5-HOMERUN	5-strikeout
6-WALK	6-HOMERUN	6-strikeout
7-WALK	7-HOMERUN	7-strikeout
8-WALK	8-HOMERUN	8-strikeout
9-WALK	1-10 flyball (lf)B 11-20	9-strikeout
10-flyball (cf)B	9-WALK	10-strikeout
11-groundball (ss)A++	10-DOUBLE** 1-11 SINGLE** 12-20	11-WALK
12-flyball (lf)B	11-SINGLE* 1-6 lineout (3b) 7-20	12-flyball (cf)A plus injury
	12-WALK	

GREG MADDUX pitcher-1 batting # 4 starter

PITCHING CARD ATLANTA

4	5	6
2-lineout (ss)	2-flyball (lf)C	2-strikeout
3-groundball (1b)C	3-FLYBALL (lf)X	3-GROUND-BALL(p)X
4-FLYBALL (cf)X	4-GROUND-BALL(ss)X	4-flyball (lf)C
5-GROUND-BALL(ss)X	5-strikeout	5-HOMERUN 1 flyball (lf)B 2-20
6-popout (3b)	6-strikeout	6-SINGLE
7-GROUND-BALL(2b)X	7-DOUBLE** 1-7 SINGLE** 8-20	7-SINGLE* 1-12 lineout (2b) 13-20
8-flyball (rf)B	8-flyball (cf)B	8-strikeout
9-strikeout	9-strikeout	9-strikeout
10-GROUND-BALL(3b)X	10-CATCHER'S CARD X	10-groundball (2b)C
11-FLYBALL (rf)X	11-GROUND-BALL(1b)X	11-groundball (p)B
12-flyball (rf)B	12-groundball (2b)C	12-lineout (ss)

FIGURE 5.5
Mark McGwire and Greg Maddux *Strat-O-Matic* cards from the 1998 baseball seasons.

Let's illustrate playing this game for two plate appearances.

1. For the first plate appearance, we roll a 2 on the white die and so we refer to the "2" column of McGwire's card. We roll the two red dice and get a 2 and 3 for a sum of 5. We look at the number "5" in the "2" column and read the result—HOME RUN! Mac has hit a home run against Greg Maddux.

2. For the second plate appearance, we roll a 5 on the white die and we refer to the "5" column of Maddux's card. The roll of the red dice is 5 and 2 for a sum of 7.

Looking at the "7" line, we see that the result is DOUBLE if the roll of the twenty-sided die is between 1–7 and SINGLE if the die roll is between 8 and 20. We roll the twenty-sided die and get a 10—McGwire has hit a single against Maddux.

To really get an understanding how the game works, we need to calculate probabilities for the sum of two dice. As in Case Study 5-2, we distinguish the two red dice—the 36 possible outcomes are displayed in Table 5.7.

TABLE 5.7
Outcomes of the rolls of two dice.

		2nd die					
		1	2	3	4	5	6
	1	x	x	x	x	x	x
	2	x	x	x	x	x	x
1st	3	x	x	x	x	x	x
die	4	x	x	x	x	x	x
	5	x	x	x	x	x	x
	6	x	x	x	x	x	x

Since each possible outcome has the same chance, we assign a probability of $\frac{1}{36}$ to each outcome. So

$$\text{Prob(1st die is 4, 2nd die is 3)} = \tfrac{1}{36}, \quad \text{Prob(1st die is 5, 2nd die is 6)} = \tfrac{1}{36}.$$

We are interested in probabilities about the sum of the two dice.

First, we think of possible values of the sum. Table 5.8 shows the value for the sum for each possible rolls of the two dice. Looking at the table, we see that the possible sums are

$$2, 3, 4, \ldots, 12.$$

We find the probabilities of the different sums by adding the probabilities of the individual outcomes for the two dice. For example, suppose we wish to compute the probability

TABLE 5.8
Table of the sum of the rolls of two dice for all possible outcomes.

		2nd die					
		1	2	3	4	5	6
	1	2	3	4	5	6	7
	2	3	4	5	6	7	8
1st	3	4	5	6	7	8	9
die	4	5	6	7	8	9	10
	5	6	7	8	9	10	11
	6	7	8	9	10	11	12

that the sum is 4. We first note that we can get a sum of 4 three ways:

(1st die is 1, 2nd die is 3), (1st die is 2, 2nd die is 2), (1st die is 3, 2nd die is 1)

and since the probability of each outcome is $\frac{1}{36}$,

$$\text{Prob(sum is 4)} = \frac{1}{36} + \frac{1}{36} + \frac{1}{36} = \frac{3}{16}.$$

If we continue this way, we obtain the probability table for the sum shown in Table 5.9.

TABLE 5.9
Probability distribution for the sum of two fair dice.

Sum	Probability
2	$\frac{1}{36}$
3	$\frac{2}{36}$
4	$\frac{3}{36}$
5	$\frac{4}{36}$
6	$\frac{5}{36}$
7	$\frac{6}{36}$
8	$\frac{5}{36}$
9	$\frac{4}{36}$
10	$\frac{3}{36}$
11	$\frac{2}{36}$
12	$\frac{1}{36}$
Total	1

Now we can compute some probabilities of the *Strat-O-Matic* game. Look at McGwire's card. If we roll a 1 on the white die, what is the probability that he walks? From his card, we see that he walks if the sum of the dice is 5, 6, 7, 8, 9—so

$$\text{Prob(walk if we roll a 1)} = \text{Prob(sum is 5)} + \text{Prob(sum is 6)}$$
$$+ \cdots + \text{Prob(sum is 9)} = \frac{24}{36}.$$

If we roll a 3 on the white die, what is the probability that he strikes out?

$$\text{Prob(strikeout if we roll a 3)} = \text{Prob(sum is 4 through 10)} = \frac{30}{36}.$$

We use similar logic for finding probabilities off of Maddux's card. If we roll a 4 on the white die (look at Maddux's card), we see that a fly ball results if we roll a sum equal to 4, 8, 11. So

Prob(flyball if we roll a 4) = Prob(sum is 4) + Prob(sum is 8) + Prob(sum is 11) = $\frac{10}{36}$.

Actually we are interested in

Prob(Mac hits a home run)

if Mac is facing Greg Maddux. Figure 5.6 shows, using what is commonly called a *tree diagram*, all of the ways McGwire can hit a home run off Maddux.

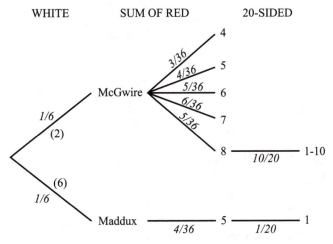

FIGURE 5.6
Tree diagram to illustrate the computation that Mark McGwire hits a home run against Greg Maddux.

- He can hit a home run if one rolls a 2 on the white die (look at column 2 of McGwire's card), and rolls a sum of 4, 5, 6, 7, or 8 on the red dice. If one rolls an 8, then one uses the 20-sided die and a roll between 1 and 10 results in a home run.

- Mac also hits a home run if one rolls 6 on the white die (look at column 6 of Maddux's card), rolls a sum of 5 on the red dice, and rolls a 1 on the 20-sided die.

In the diagram, the branches of the tree are labeled with the corresponding probabilities.

By use of the multiplication rule, we multiply the probabilities across a particular branch to find the probability of a particular outcome of the white, red, and twenty-sided dice. For example, the probability

Prob(White die is 6 AND the sum of the red dice is 5 AND the twenty-sided die is 1)

is found by multiplying the conditional probabilities

Prob(White die is 6) × Prob(Sum of red dice is 5 IF white die is 6)

× Prob(Twenty-sided die is 1 IF white die is 6 and sum of red dice is 5)

$$= \left(\tfrac{1}{6}\right) \times \left(\tfrac{4}{36}\right) \times \left(\tfrac{1}{20}\right).$$

We find the probability of Mac hitting a home run off Greg Maddux by multiplying the probabilities along each branch for each possible way of Mac hitting a home run, and adding the products. We then find the probability to be

$$\text{Prob(home run)} = \tfrac{1}{6} \times \tfrac{3}{36}$$
$$+ \tfrac{1}{6} \times \tfrac{4}{36}$$
$$+ \tfrac{1}{6} \times \tfrac{5}{36}$$
$$+ \tfrac{1}{6} \times \tfrac{6}{36}$$
$$+ \tfrac{1}{6} \times \tfrac{5}{36} \times \tfrac{10}{20}$$
$$+ \tfrac{1}{6} \times \tfrac{4}{36} \times \tfrac{1}{20}$$
$$= 0.0958.$$

Exercises

Leadoff Exercise. Table 5.10 shows the basic batting statistics for Rickey Henderson for the 1990 season.

TABLE 5.10
Batting statistics for Rickey Henderson for the 1990 season.

AB	R	H	2B	3B	HR	RBI	BB	SO
489	119	159	33	3	28	61	97	60

Construct a random spinner for Rickey using his 1990 season. Compute the number of plate appearances in the 1990 season. Find the approximate probability that Rickey gets (1) a single, (2) a double, (3) a triple, (4) a home run, (5) a walk, and (6) an out. Make a spinner like the one described in Case Study 5-3 where the areas of the regions correspond to the probabilities of the batting events that you computed.

5.1. Table 5.11 shows the number of at-bats (AB) and hits (H) for Phillie Doug Glanville for every game that he played during the 1999 season. The table also has two more columns:

- CAB (for cumulative at-bats), the number of at-bats on that day and all earlier days.

- CH (for cumulative hits), the number of hits on that day and all earlier days.

(a) Compute Glanville's batting average after playing five games. Do you think this is a good estimate of Glanville's ability to get a base hit?

TABLE 5.11
Number of at-bats and hits for Doug Glanville for all games played during the 1999 season.

Date	AB	H	CAB	CH	Date	AB	H	CAB	CH	Date	AB	H	CAB	CH
04/05/99	3	1	3	1	06/04/99	6	4	205	63	08/03/99	5	2	430	140
04/06/99	4	1	7	2	06/05/99	5	2	210	65	08/04/99	4	3	434	143
04/07/99	4	1	11	3	06/06/99	5	2	215	67	08/05/99	4	2	438	145
04/08/99	3	1	14	4	06/07/99	4	2	219	69	08/06/99	5	1	443	146
04/09/99	3	0	17	4	06/08/99	4	1	223	70	08/07/99	5	1	448	147
04/10/99	4	0	21	4	06/09/99	5	3	228	73	08/08/99	0	0	448	147
04/11/99	0	0	21	4	06/11/99	4	1	232	74	08/09/99	5	1	453	148
04/12/99	5	2	26	6	06/12/99	5	2	237	76	08/10/99	4	1	457	149
04/14/99	5	1	31	7	06/13/99	4	0	241	76	08/11/99	4	1	461	150
04/16/99	6	3	37	10	06/15/99	5	1	246	77	08/13/99	5	2	466	152
04/17/99	5	1	42	11	06/16/99	5	2	251	79	08/14/99	4	2	470	154
04/18/99	5	3	47	14	06/17/99	5	2	256	81	08/15/99	3	0	473	154
04/19/99	4	0	51	14	06/18/99	4	1	260	82	08/17/99	5	3	478	157
04/20/99	4	3	55	17	06/19/99	4	1	264	83	08/18/99	4	2	482	159
04/21/99	3	1	58	18	06/20/99	3	1	267	84	08/20/99	4	2	486	161
04/23/99	3	1	61	19	06/22/99	4	1	271	85	08/21/99	4	1	490	162
04/24/99	4	1	65	20	06/23/99	3	0	274	85	08/22/99	0	0	490	162
04/25/99	4	1	69	21	06/24/99	5	1	279	86	08/28/99	5	1	495	163
04/27/99	5	3	74	24	06/25/99	4	2	283	88	08/29/99	5	3	500	166
04/28/99	4	1	78	25	06/27/99	5	2	288	90	08/30/99	5	1	505	167
04/29/99	4	0	82	25	06/28/99	4	0	292	90	**08/31/99**	**4**	**0**	**509**	**167**
04/30/99	**5**	**1**	**87**	**26**	06/29/99	4	2	296	92	09/01/99	5	1	514	168
05/01/99	4	1	91	27	**06/30/99**	**5**	**2**	**301**	**94**	09/02/99	4	2	518	170
05/02/99	4	2	95	29	07/01/99	4	1	305	95	09/03/99	4	1	522	171
05/03/99	4	0	99	29	07/02/99	5	3	310	98	09/04/99	3	2	525	173
05/04/99	4	1	103	30	07/03/99	5	3	315	101	09/05/99	4	0	529	173
05/07/99	3	2	106	32	07/04/99	5	1	320	102	09/06/99	5	1	534	174
05/08/99	3	2	109	34	07/05/99	5	2	325	104	09/07/99	5	2	539	176
05/09/99	4	1	113	35	07/07/99	5	2	330	106	09/08/99	3	0	542	176
05/10/99	5	2	118	37	07/09/99	5	2	335	108	09/09/99	1	0	543	176
05/11/99	3	1	121	38	07/10/99	4	0	339	108	09/10/99	3	0	546	176
05/12/99	4	2	125	40	07/11/99	4	2	343	110	09/11/99	4	3	550	179
05/14/99	4	0	129	40	07/15/99	5	3	348	113	09/12/99	5	1	555	180
05/15/99	5	3	134	43	07/16/99	4	2	352	115	09/13/99	3	0	558	180
05/16/99	3	1	137	44	07/17/99	6	4	358	119	09/14/99	4	1	562	181
05/17/99	5	1	142	45	07/18/99	5	1	363	120	09/15/99	6	5	568	186

(continued)

TABLE 5.11
Continued.

Date	AB	H	CAB	CH	Date	AB	H	CAB	CH	Date	AB	H	CAB	CH
05/18/99	5	1	147	46	07/19/99	5	2	368	122	09/17/99	4	1	572	187
05/19/99	1	0	148	46	07/20/99	6	3	374	125	09/18/99	4	2	576	189
05/21/99	5	3	153	49	07/21/99	5	3	379	128	09/19/99	4	1	580	190
05/22/99	4	2	157	51	07/22/99	4	0	383	128	09/20/99	5	2	585	192
05/23/99	4	2	161	53	07/23/99	4	1	387	129	09/22/99	5	2	590	194
05/24/99	4	1	165	54	07/23/99	4	2	391	131	09/23/99	5	1	595	195
05/25/99	5	1	170	55	07/24/99	4	2	395	133	09/24/99	4	1	599	196
05/26/99	3	0	173	55	07/25/99	4	0	399	133	09/25/99	3	0	602	196
05/28/99	5	1	178	56	07/27/99	4	0	403	133	09/26/99	4	1	606	197
05/29/99	3	1	181	57	07/28/99	4	1	407	134	09/28/99	4	2	610	199
05/30/99	3	0	184	57	07/29/99	5	3	412	137	**09/29/99**	**5**	**2**	**615**	**201**
05/31/99	**4**	**0**	**188**	**57**	07/30/99	5	0	417	137	10/01/99	5	2	620	203
06/01/99	6	2	194	59	**07/31/99**	**4**	**0**	**421**	**137**	10/02/99	5	1	625	204
06/03/99	5	0	199	59	08/01/99	4	1	425	138	10/03/99	3	0	628	204

(b) Compute Glanville's batting average after playing ten games. Do you think this average is a better estimate of Glanville's "true" batting average than the value you computed in part (a)? Why?

(c) Compute Glanville's batting average at the end of each month and at the end of the baseball season. (These dates are in bold-face type above.) Put your batting averages in the table below.

Date	CAB	CB	AVG
end of April			
end of May			
end of June			
end of July			
end of August			
end of September			
end of season			

(d) Plot the batting averages against the month in the grid below.

(e) What pattern do you see in your graph? What do you think is Glanville's true batting average? Why?

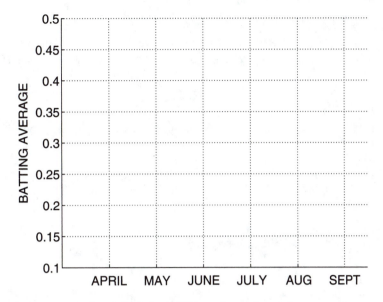

5.2. Kenny Lofton, the Indians' 1999 leadoff hitter, is coming to bat.

    (a) Without using any data, guess at the probability that Lofton will get a triple on this at-bat.

    (b) Table 5.12 shows the number of at-bats (AB) and the number of triples (3B) for Lofton after particular dates during the 1999 season. For each date, compute the proportion of triples. Put your answers in the Proportion column of the table.

TABLE 5.12
Number of at-bats and triples for Kenny Lofton after particular dates during the 1999 season.

Date	AB	3B	Proportion
04/21/99	52	2	
05/03/99	103	3	
05/19/99	152	3	
06/05/99	202	3	
06/19/99	250	4	
07/03/99	303	4	
07/20/99	354	4	
09/15/99	403	5	
09/29/99	452	5	

    (c) Based on your calculations, what is your best estimate of the probability that Lofton will get a triple? Explain why this is your best guess.

(d) Suppose a hitter gets no triples in 400 at-bats. Does this mean that the probability that he hits a triple is exactly zero? Why or why not?

5.3. If you had watched Mark McGwire bat, you would have noticed that he tends to strike out a lot. After each game of the 1999 season, I computed the strikeout (SO) rate

$$\text{SO Rate} = \frac{\text{SO}}{\text{AB}}.$$

Figure 5.7 graphs McGwire's strikeout rate after each game of the 1999 season.

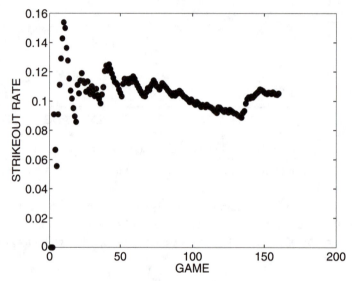

FIGURE 5.7
Plot of strikeout rate against game number of Mark McGwire for the 1999 baseball season.

(a) Using the graph, estimate McGwire's probability of striking out in an at-bat after 50 games.

(b) Estimate the chance of striking out in an at-bat after 100 games.

(c) What is your best estimate at McGwire's strikeout probability based on this data? Use the graph.

5.4. (Exercise 1.24, continued) Table 5.13 shows the number of innings pitched (IP) and the number of earned runs (ER) for each game that Charles Nagy started during the 1999 season. The CIP column gives the number of innings pitched in all games in that game and earlier, and CER gives the number of earned runs in that game and earlier.

TABLE 5.13

Number of innings pitched and earned runs for all of the games Charles Nagy started in the 1999 season.

Game	Date	IP	ER	CIP	CER	Game	Date	IP	ER	CIP	CER
1	04/09/99	7	3	7	3	18	07/17/99	4	4	107.67	55
2	04/17/99	7.67	1	14.67	4	19	07/22/99	6	4	113.67	59
3	04/22/99	6.67	4	21.33	8	20	07/27/99	8	5	121.67	64
4	04/27/99	6.67	3	28	11	21	08/01/99	8	5	129.67	69
5	05/02/99	3.33	8	31.33	19	22	08/07/99	5	7	134.67	76
6	05/08/99	3.67	5	35	24	23	08/13/99	7	1	141.67	77
7	05/14/99	6.67	2	41.67	26	24	08/18/99	6	5	147.67	82
8	05/19/99	8	1	49.67	27	25	08/23/99	8	4	155.67	86
9	05/25/99	7	1	56.67	28	26	08/28/99	8	0	163.67	86
10	05/31/99	6	1	62.67	29	27	09/02/99	7	1	170.67	87
11	06/06/99	6	2	68.67	31	28	09/07/99	6	2	176.67	89
12	06/12/99	6.33	3	75	34	29	09/13/99	5.67	5	182.33	94
13	06/18/99	5.67	3	80.67	37	30	09/18/99	6.67	3	189	97
14	06/23/99	7	4	87.67	41	31	09/23/99	5	6	194	103
15	06/28/99	7	1	94.67	42	32	09/29/99	7	5	201	108
16	07/03/99	2	8	96.67	50	33	10/03/99	1	3	202	111
17	07/08/99	7	1	103.67	51						

(a) Find Nagy's earned run average (ERA) after the first game in the 1999 season.

(b) Find Nagy's ERA after ten games, twenty games, and at the end of the season. Put your answers in the table below.

Game	ERA
1	
10	
20	
33	

(c) Figure 5.8 plots Nagy's ERA after each game pitched in the 1999 season.

(d) Describe the pattern of the graph from left to right. What does this pattern mean in terms of Nagy's pitching performance during the 1999 season?

(e) Suppose that Nagy pitched one more inning in 1999 and allowed eight runs. Without calculating anything, do you think Nagy's ERA would go up? Now calculate what the ERA would be. Is this more or less than you expected?

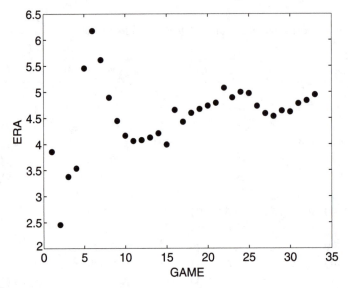

FIGURE 5.8
Plot of season ERA against game number for Charles Nagy after each game he pitched in the 1999 season.

5.5. In Case Study 5-2, a basic tabletop baseball game, *Big League Baseball*, is described. This game was played on a computer for 1000 games. The number of home runs hit in each game was recorded and these data are summarized in Table 5.14.

TABLE 5.14
Frequency table of the number of home runs hit in 1000 simulated games of *Big League Baseball*.

Number of home runs hit	0	1	2	3	4	5	6	7
Count	165	289	263	170	75	29	5	4

(a) Find the probability that no home runs are hit in a game.

(b) Find the probability that between two and four home runs are hit.

(c) Find the probability that at least one home run is hit.

(d) If you play a game of *Big League Baseball*, what is the most likely number of home runs that will be hit?

5.6. (Exercise 5.5 continued)   For each of the 1000 games played of *Big League Baseball*, the number of pitches was recorded. Table 5.15 gives a grouped frequency table of these data.

TABLE 5.15
Grouped frequency table of the number of pitches thrown in 1000 simulated games of *Big League Baseball*.

Number of pitches in a game	1–150	151–200	201–250	251–300	301–350	351–400	401–450
Count:	20	661	278	29	8	3	1

    (a) What is the probability that between 151 and 200 pitches are thrown in a game?

    (b) What is the probability that at most 250 pitches are thrown?

    (c) What would be a typical number of pitches thrown in a game?

    (d) Would you be surprised to see a game with over 300 pitches thrown? Why?

    (e) Can you think of circumstances where over 300 pitches are likely to to occur?

5.7. (Exercise 5.5 continued) For each of the 1000 games played of *Big League Baseball*, the margin of victory (winning team score − losing team score) was recorded. A frequency table of these data is given in Table 5.16.

TABLE 5.16
Frequency table of the margin of victory in 1000 simulated games of *Big League Baseball*.

Margin of victory	1	2	3	4	5	6	7	8	9	10 or more
Count	333	209	142	108	66	48	33	21	21	19

    (a) What is the probability that a game of *Big League Baseball* is decided by one run?

    (b) Suppose you define a "blowout" as a game where a team wins by six runs or more. What is the probability a game is a blowout?

    (c) What is the probability a game is not a blowout?

    (d) Is it unusual that a team would win by ten or more runs? Why?

5.8. (Exercise 5.5 continued) Suppose you are interested in exploring the relationship between the number of runners to reach base and the runs scored in a half-inning of baseball. For 1000 games of *Big League Baseball*, we record for each half-inning

- Runners—the total number of runners to reach base in the half-inning,
- Runs—the number of runs scored.

    Table 5.17 classifies 18,200 half-innings with respect to Runners and Runs.

    (a) Find the probability that there are no runners on base in a particular half-inning.

    (b) Find the probability that a team doesn't score any runs during their at-bat.

    (c) Suppose that a team has one runner on base during their half-inning—what is the probability that the runner scores?

TABLE 5.17
Two-way count table of the number of runners on base and the runs scored in half innings from 1000 simulated games of *Big League Baseball*.

Runners	Runs 0	1	2	3	4	5	6	≥ 7	Sum
0	5481	0	0	0	0	0	0	0	5481
1	5441	539	0	0	0	0	0	0	5980
2	2408	896	244	0	0	0	0	0	3548
3	463	778	434	108	0	0	0	0	1783
4	13	215	369	176	52	0	0	0	825
5	0	3	81	159	71	19	0	0	333
6	0	0	1	35	66	35	11	0	148
7	0	0	0	0	15	32	14	2	63
8	0	0	0	0	0	6	11	6	23
9	0	0	0	0	0	0	3	4	7
10	0	0	0	0	0	0	0	6	6
11	0	0	0	0	0	0	0	3	3
Sum	13806	2431	1129	478	204	92	39	21	18200

(d) Suppose that a team has three runners on base—find the probability that at least one run is scored.

5.9. (Exercise 5.8 continued)   For ten consecutive games played by the Phillies in 1999, the number of runners and the number of runs scored are recorded for each half-inning. Table 5.18 classifies all half-innings by Runs and Runners.

TABLE 5.18
Two-way count table of the number of runners on base and the runs scored in half innings from ten consecutive Phillies games in 1999.

Runners	Runs 0	1	2	3	4	5	8	Sum
0	57	0	0	0	0	0	0	57
1	46	5	0	0	0	0	0	51
2	17	12	5	0	0	0	0	34
3	3	5	9	0	0	0	0	17
4	0	1	6	1	0	0	0	8
5	0	0	1	1	1	2	0	5
6	0	0	0	1	0	1	0	2
7	0	0	0	0	1	0	0	1
10	0	0	0	0	0	0	1	1
Sum	123	23	21	3	2	3	1	176

(a) Compute the same probabilities asked in parts (a)–(d) of Exercise 5.8.

(b) Compare your answers with those of Exercise 5.8. Are the results from the game *Big League Baseball* similar to those from real baseball?

5.10. In the game *Big League Baseball* described in Case Study 5-2, a red die is thrown to represent the pitch. The possible rolls of the die and the pitch results are shown in the table below.

Red die roll	Pitch result
1, 6	batter hits fair ball
2, 3	ball
4, 5	strike

(a) What is the probability that a pitch results in a strike?

(b) What is the probability that the pitch results in the batter hitting a fair ball?

(c) What is the probability that the pitch results in a ball or a strike?

5.11. (*Big League Baseball*, continued)   Suppose two pitches are thrown to a batter. The following tree diagram shows the possible outcomes, where a ball is denoted by B and a strike by S.

1ST PITCH   2ND PITCH

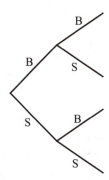

(a) Assign probabilities to each branch of the tree using the table of dice rolls in Exercise 5.10.

(b) Find the probability that two balls are thrown in a row. (To find a probability of "Ball on first pitch and Ball on second pitch" or "Ball 1, Ball 2," you multiply the probabilities along the branches of the tree.)

(c) Find the probability that exactly one of the two pitches is a strike. (Find the probability of "Ball 1, Strike 2" and "Strike 1, Ball 2" and then add the probabilities of the two ways of getting one strike.)

(d) Find the probability that at least one of the pitches is a strike.

5.12. (*Big League Baseball*, continued)   Use a tree diagram like the one in Exercise 5.11 to answer the questions about what happens on three or more pitches.

(a) Find the probability that a batter strikes out on three consecutive pitches.

(b) Suppose the pitch count is 2–1 (that is, 2 balls and 1 strike) after three pitches. List below all the possible pitch sequences (like BSB) which would result in a 2–1 count.

(c) Find the probability that the pitch count is 2–1 after three pitches. (*Hint:* Find the probability of each pitch sequence in (b) and then add the probabilities of all the sequences to obtain the probability you want.)

(d) Find the probability that a batter strikes out on four pitches.

(e) Find the probability that a batter strikes out on at most four pitches.

5.13. (*Big League Baseball*, continued) If the roll of the red die is 1 or 6, then the batter hits a fair ball. Two white dice are rolled and the outcome of the play depends on the roll of the dice as shown in Table 5.19.

TABLE 5.19
Play outcomes from the rolls of two dice from *Big League Baseball*.

White Die 1	White Die 2					
	1	2	3	4	5	6
1	Single	Out	Out	Out	Out	Error
2	Out	Double	Single	Out	Single	Out
3	Out	Single	Triple	Out	Out	Out
4	Out	Out	Out	Out	Out	Out
5	Out	Single	Out	Out	Out	Single
6	Error	Out	Out	Out	Single	Home run

Using the table,

(a) Find the probability that the batter gets a double.

(b) Find the probability that the batter gets a triple.

(c) Find the probability the batter gets a hit (single, double, triple, or home run).

(d) Find the probability that the batter gets on base. (Note that getting on base is different from getting a hit.)

(e) Find the probability that the batter gets on base as a result of an error.

5.14. (*Big League Baseball*, continued) In the roll of the white dice, suppose that we consider *only* the outcomes that result in hits. The results of the rolls of the white dice that result in hits are shown in Table 5.20; all other results are left blank. Suppose that each hit shown in the table has the same probability.

(a) How many outcomes of the two dice result in a hit?

(b) Find the probability that a hit is a single.

TABLE 5.20

Play outcomes from the rolls of two dice from *Big League Baseball* that result in hits.

White Die 1	White Die 2					
	1	2	3	4	5	6
1	Single					
2		Double	Single		Single	
3		Single	Triple			
4						
5		Single				Single
6					Single	HR

(c) Find the probability that a hit is a double.

(d) Find the probability that a hit is *not* a single.

5.15. Table 5.21 gives total at-bats, singles, doubles, etc. for all Major League games played in the 1999 season.

TABLE 5.21

Total offensive statistics for all games played in the 1999 season.

AB	1B	2B	3B	HR	BB	SO	HBP
167137	30128	8740	931	5528	17891	31119	1579

(a) Compute the number of plate appearances (PA) for the 1999 season.

(b) Find the probability that a hitter gets a single in a PA.

(c) Find the probability the hitter gets an extra base hit in a PA.

(d) In a PA, what is the most likely outcome: a hit, a strikeout, a walk, or an out? Explain.

5.16. (Exercise 5.15, continued)

(a) Consider only the PAs that get on base (including home runs) in the 1999 offensive data. In the table below, put the number of hits of different types and the number of walks and HBPs, and find the proportion of each type.

(b) Find the probability that an on-base event is for extra bases (an extra base is a hit for more than one base; it includes doubles, triples, and home runs).

(c) Find the probability that a person on-base has reached there by a walk or an HBP.

(d) Compare your proportions in the table above with the on-base profile from the tabletop game *Big League Baseball*. Are there any similarities? Any differences?

On-base Profile

Type	Count	Proportion
Single		
Double		
Triple		
Home run		
Walk		
HBP		
TOTAL		

5.17. (*All-Star Baseball*) The spinner shown in Figure 5.9 was created using Ty Cobb's batting statistics for the year 1911. As in Case Study 5-3, the spinner has 36 areas of the same size. By spinning the spinner, one is simulating the result of a single plate appearance by Ty Cobb during the 1911 season.

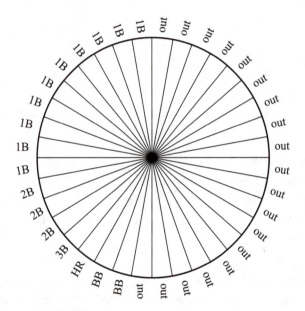

FIGURE 5.9
Spinner for Ty Cobb based on 1911 batting statistics.

(a) Find the probability that Cobb gets a home run on a single plate appearance.

(b) Find the probability that Cobb gets a single.

(c) Find the probability that Cobb gets on-base.

(d) Find the most likely outcome for Cobb on a single plate appearance.

(e) Find the probability that Cobb gets an extra base hit.

5.18. (Creating a random spinner for *All-Star Baseball* for a player from a season of batting data)   Table 5.22 gives basic hitting statistics for Mickey Mantle for the 1956 baseball season. For simplicity, we assume that plate appearances (PA) are recorded as either at-bats (AB) or walks (BB). (Hit by pitches and sacrifice hits are ignored) Also we assume at-bats that are not hits are either strikeouts (SO), or groundouts or flyouts.

TABLE 5.22

Mickey Mantle's 1956 batting statistics.

	PA	AB	H	1B	2B	3B	HR	BB	SO	Groundouts and Flyouts
Count		533	188		22	5	52	99	112	
Probability	xxxx	xxxx	xxxx							
Regions = 36 × Prob	xxxx	xxxx	xxxx							

(a) Compute the number of Plate Appearances (PA) by adding the AB and BB.

(b) Compute the number of singles (1B) by adding up doubles (2B), triples (3B), and home runs (HR), and subtracting this sum from the number of hits (H).

(c) Find the number of groundouts or flyouts by adding H and SO, and subtracting this sum from the number of at-bats (AB).

(d) Find the proportion of PAs that are 1B, 2B, 3B, HR, BB, SO, and Groundouts or Flyouts-put the results in the probability row.

(e) Convert the probabilities to spinner regions by multiplying by 36 and rounding to the nearest whole number as was done in Case Study 5-3.

(f) Construct the random spinner (from the blank one shown below) using the region values computed in part (e).

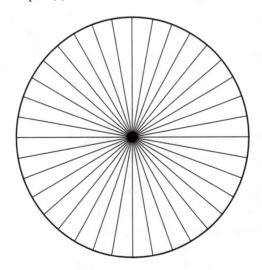

5.19. (Group activity)   Suppose you are interested in playing a game of *All-Star Baseball* between the greatest players of the National League (NL) and the greatest players in the American League (AL). Using the career statistics for the 18 great players given in Table 5.23, construct a set of random spinners using the same method as Exercise 5.18.

TABLE 5.23
Career batting statistics for 18 great players.

Catcher

	G	AB	R	H	2B	3B	HR	SO	BB	AVG	OBP	SLG
(NL) Johnny Bench	2158	7658	1091	2048	381	24	389	1278	891	.267	.342	.476
(AL) Yogi Berra	2120	7555	1175	2150	321	49	358	414	704	.285	.348	.482

First Base

	G	AB	R	H	2B	3B	HR	SO	BB	AVG	OBP	SLG
(AL) Lou Gehrig	2164	8001	1888	2721	534	163	493	790	1508	.340	.447	.632
(NL) Mark McGwire	1745	5830	1109	1553	247	6	547	1461	1247	.266	.397	.592

Second Base

	G	AB	R	H	2B	3B	HR	SO	BB	AVG	OBP	SLG
(NL) Jackie Robinson	1382	4877	947	1518	273	54	137	291	740	.311	.409	.474
(AL) Eddie Collins	2826	9949	1821	3315	438	187	47	286	1499	.333	.424	.429

Shortstop

	G	AB	R	H	2B	3B	HR	SO	BB	AVG	OBP	SLG
(AL) Cal Ripken, Jr.	2848	10982	1592	3046	580	44	415	1228	1095	.277	.343	.452
(NL) Honus Wagner	2792	10430	1736	3415	640	252	101	327	963	.327	.391	.466

Third Base

	G	AB	R	H	2B	3B	HR	SO	BB	AVG	OBP	SLG
(NL) Mike Schmidt	2404	8352	1506	2234	408	59	548	1883	1507	.267	.380	.527
(AL) Brooks Robinson	2896	10654	1232	2848	482	68	268	990	886	.267	.322	.401

Outfielders

	G	AB	R	H	2B	3B	HR	SO	BB	AVG	OBP	SLG
(AL) Babe Ruth	2503	8399	2174	2873	506	136	714	1330	2062	.342	.474	.690
(AL) Ted Williams	2292	7706	1798	2654	525	71	521	709	2019	.344	.482	.634
(NL) Willie Mays	2992	10881	2062	3283	523	140	660	1526	1464	.302	.384	.557
(NL) Hank Aaron	3298	12364	2174	3771	624	98	755	1383	1402	.305	.374	.555
(AL) Joe DiMaggio	1736	6821	1390	2214	389	131	361	369	790	.325	.398	.579
(AL) Mickey Mantle	2401	8102	1677	2415	344	72	536	1710	1733	.298	.421	.557
(NL) Pete Rose	3562	14053	2165	4256	746	135	160	1143	1566	.303	.375	.409
(NL) Stan Musial	3026	10972	1949	3630	725	177	475	696	1599	.331	.417	.559

5.20. (Probabilities of sum of two dice) Suppose you toss two dice. Assume that the dice are distinguishable (think of one die as white and the other die as red) and there are 36 possible outcomes shown in Table 5.24. The outcomes are written in the table as "roll of white, roll of red." So, for example, a "4, 3" indicates that the roll of the white die was 4 and the roll of the red die was 3, a "3, 5" indicates that the white die was 3 and the red die was 5, and so on.

TABLE 5.24
Outcomes of rolling a white die and a red die.

Roll of white die	Roll of red die					
	1	2	3	4	5	6
1	1, 1	1, 2	1, 3	1, 4	1, 5	1, 6
2	2, 1	2, 2	2, 3	2, 4	2, 5	2, 6
3	3, 1	3, 2	3, 3	3, 4	3, 5	3, 6
4	4, 1	4, 2	4, 3	4, 4	4, 5	4, 6
5	5, 1	5, 2	5, 3	5, 4	5, 5	5, 6
6	6, 1	6, 2	6, 3	6, 4	6, 5	6, 6

(a) List the outcomes in the table where the roll of the white die is equal to 2. (One of these outcomes is "2, 3.")

(b) List the outcomes where the roll of the white die is equal to the roll on the red die.

(c) List the outcomes where the roll of the white die is greater than the roll on the red die.

(d) Find the probabilities of the events

  i. "roll of white die is equal to 2,"

  ii. "roll of white die is equal to roll of red die,"

  iii. "roll of white die is greater than roll of red die."

5.21. (*Strat-O-Matic* baseball) Suppose the 1998 Tony Gwynn is facing the 1998 Randy Johnson. (A great hitter against a great pitcher.) Three dice, one red and two white, are tossed to determine the result. The red die determines which card to look at: If the red die is 1, 2, 3, Gwynn's card is used and if the die is 4, 5, 6, Johnson's card is used (these numbers are shown at the top of the cards). The sum of the two white dice determines which number is checked under the red dice number. Suppose that the roll of the red die is 3 and the sum of the white dice is 9. We check the number 9 under Gwynn's 3 column and read that Gwynn has walked on this particular at-bat.

(a) What is the chance that the Gwynn card is used (when the red die is rolled 1, 2, or 3)?

(b) Suppose that the roll of the red die is 2 (so the middle column of Gwynn's card is used). What rolls of the sum of the white dice will result in a groundball?

TONY GWYNN rightfield-3	stealing-D running 1-10	**RANDY JOHNSON**	pitcher-3 batting # 1	starter

SAN DIEGO **PITCHING CARD** **SEATTLE**

1	2	3	4	5	6
2-lineout (1b) into as many outs as possible	2-lineout (1b)	2-lineout (2b)	2-FLYBALL (lf)X	2-strikeout	2-GROUND-BALL(ss)X
3-groundball (2b)A++ plus injury	3-groundball (3b)B	3-groundball (3b)A	3-GROUND-BALL(1b)X	3-WALK	3-FLYBALL (rf)X
4-HOMERUN	4-groundball (1b)A	4-flyball (rf)B	4-CATCHER'S CARD X	4-strikeout	4-GROUND-BALL(2b)X
5-HOMERUN 1-6	5-flyball (cf)B	5-SINGLE**	5-strikeout	5-strikeout	5-strikeout
DOUBLE 7-20	6-popout (2b)	6-SINGLE**	6-strikeout	6-DOUBLE** 1-13	6-strikeout
6-DOUBLE	7-popout (1b)	7-groundball (1b)A	7-WALK	SINGLE** 14-20	7-SINGLE
7-DOUBLE** 1-7	8-lineout (2b)	8-groundball (ss)A	8-strikeout	7-GROUND-BALL(ss)X	8-strikeout
SINGLE** 8-20	9-flyball (rf)B	9-WALK	9-strikeout	8-strikeout	9-strikeout
8-SINGLE	10-groundball (p)A	10-SINGLE* 1-4	10-GROUND-BALL(2b)X	9-HOMERUN 1-15	10-GROUND-BALL(3b)X
9-groundball (2b)A++	11-groundball (3b)A	lineout (2b) 5-20	11-strikeout	DOUBLE 16-20	11-GROUND-BALL(p)X
10-SINGLE*	12-lineout (2b)	11-groundball (2b)A	12-FLYBALL (lf)X	10-FLYBALL (cf)X	12-groundball (2b)C
11-SINGLE		12-foulout (c)		11-WALK	
12-popout (2b)				12-SINGLE* 1-12 lineout (3b) 13-20	

FIGURE 5.10

Strat-O-Matic cards for the 1998 Tony Gwynn and the 1998 Randy Johnson.

   (c) If the roll of the red die is 6, which rolls of the sum of the white dice will result in a strikeout for Gwynn?

   (d) If the roll of the red die is 1, which rolls of the sum of the white die will result in a hit?

5.22. (*Strat-O-Matic* continued) The 1998 Tony Gwynn is facing the 1998 Randy Johnson.

   (a) If the roll of the red die is 6, the table below gives the possible rolls of the sum of white dice, the probabilities, and the outcomes (copied from the Johnson card above). In this case (the roll of the red die is 6) . . .

   i. What is the probability that Gwynn will single?

   ii. What is the probability that Gwynn will strike out?

   iii. What is the probability that Gwynn will hit a ground ball?

   (b) If the red die is rolled 2, what is the probability that Gwynn will pop out? (Look at the Gwynn card above.)

Sum of white dice	Probability	Outcome
2	$\frac{1}{36}$	Groundball
3	$\frac{2}{36}$	Flyball
4	$\frac{3}{36}$	Groundball
5	$\frac{4}{36}$	Strikeout
6	$\frac{5}{36}$	Strikeout
7	$\frac{6}{36}$	Single
8	$\frac{5}{36}$	Strikeout
9	$\frac{4}{36}$	Strikeout
10	$\frac{3}{36}$	Groundball
11	$\frac{2}{36}$	Groundball
12	$\frac{1}{36}$	Groundball

5.23. (*Strat-O-Matic* continued)   The 1998 Tony Gwynn is facing the 1998 Randy Johnson. Suppose that we are interested in the probability that Gwynn strikes out against Johnson. The event that "Gwynn strikes out" can be divided into six different events.

The roll of the red die is 1 and Gwynn strikes out.

The roll of the red die is 2 and Gwynn strikes out.

The roll of the red die is 3 and Gwynn strikes out.

The roll of the red die is 4 and Gwynn strikes out.

The roll of the red die is 5 and Gwynn strikes out.

The roll of the red die is 6 and Gwynn strikes out.

These events are represented as the six branches of the tree diagram below. To find the probability that Gwynn strikes out, we first find the probabilities of all of the subbranches and then add the products of the probabilities along the subbranches to get the desired result. We outline these calculations below.

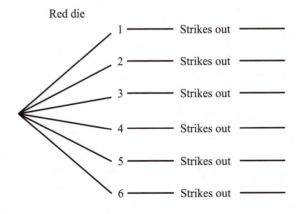

(a) First, find the probability that the roll of the red die is 1, 2, ... , 6, and place these probabilities at the six branches under the label "Red die" in the diagram.

(b) If the roll of the red die is 1 (so that you look at Gwynn's card in the "1" column), find the probability that Gwynn strikes out. Place this probability above the first horizontal line in the second set of branches.

(c) If the roll of the red die is 2, find the probability of a strikeout. Similarly, find the probability of a strikeout if the red die falls 3, 4, 5, and 6. Place these five probabilities of a strikeout above the corresponding horizontal lines to the right of 2, 3, 4, 5, 6 in the diagram.

(d) The probability of a strikeout is found in two steps:

    i. *Multiply* (for each possible red die roll) the probability of the red die roll and the probability of a strikeout for that red die roll, placing them in the blank lines to the right of the word "strikes out."

    ii. *Add* the six products you found above.

You have now found the probability of Gwynn striking out.

5.24. (*Strat-O-Matic* continued)   Use the tree diagram method described in Exercise 5.23 to find the probability that Tony Gwynn gets a walk against Randy Johnson.

Further Reading

Basic concepts of probability are presented in Devore and Peck (2000), Moore and McCabe (1998), and Scheaffer (1995). Albert and Bennett (2001), Chapter 1, discuss the probability models behind some popular tabletop baseball games, including *All-Star Baseball*, *Strat-O-Matic Baseball*, and *APBA Baseball*.

6

Probability Distributions and Baseball

What's On-Deck?

In this chapter, we show how some basic discrete probability distributions can be used to model events in baseball. One nice feature of baseball is its discrete structure. A player comes to bat during an inning, and there is a result of this plate appearance, which might be a hit, a walk, an out, an error, or a hit-by-pitch. The number of hits by a player in a given number of at-bats can be represented by a binomial distribution where the probability of a hit is the player's "true" batting average. We show in Case Study 6-1 that binomial probabilities can be very useful in predicting the number of games with 0 hits, 1 hit, and so on for a particular hitter. The remainder of the chapter focuses on modeling of run production for a team. To score runs, the batters need to get on base, and Case Study 6-2 considers a probability distribution for the number of batters that come to bat in a half-inning. If the result of each batting appearance is "out" or "get on base," and we are interested in the number of hitters B until three outs, then we can model B with a negative binomial distribution. If we can estimate a team's on-base probability, then we can find a probability distribution for the number of batters that come to bat during an inning. The second step of the run scoring process is advancing runners that get on-base. Suppose that we know the proportion of on-base events of a team that are walks, singles, doubles, triples, and home runs. (We call this set of proportions the team's on-base profile.) If we know that a team has a particular number of on-base events, then Case Study 6-3 shows how we can use a team's on-base profile to compute the probability that the team scores a given number of runs. By looking at the actual runs scored by the 1987 Chicago Cubs, we will see that this model does a reasonable job in explaining a team's run production.

Case Study 6-1: The Binomial Distribution and Hits per Game

Topics Covered: Binomial probabilities, independence, expected counts, simulation   The binomial distribution is one of the most useful probability distributions in statistics. It's a most useful distribution since it is applicable in a wide variety of situations, including baseball. Suppose we have an experiment consisting of a sequence of n identical "trials," where each trial can result in one of two possibilities, called a success and a failure. The chance of a success, p, is assumed constant for each trial, and the results of different trials are assumed independent. Then the random variable X, the number of successes in n trials, has a binomial probability function given by

$$\Pr(X = x) = \binom{n}{x} p^x (1 - p)^{n-x},$$

where $\binom{n}{x}$ is the binomial coefficient

$$\binom{n}{x} = \frac{n!}{x!(n - x)!},$$

x is an integer and the values of x can range from 0 to n.

Suppose a baseball hitter comes to bat n times during a game. For this example, we will only consider official at-bats, where the hitter gets a hit or produces an out—a walk or hit-by-pitch or a sacrifice fly are not considered official at-bats. For each at-bat, we will call a base hit a success, and an out a failure. Suppose that p, the probability of a hit, remains constant for all at-bats, and the results of different at-bats are independent. Then X, the number of hits in n at-bats during a game, will have a binomial distribution with *parameters n* and *p*.

Before we try to fit a binomial distribution to hitting data, we should ask ourselves if the above assumptions make sense. One important assumption is that the probability of a hit for a batter doesn't change across a game. In our modeling, it is convenient to go one step further and assume that the probability of a hit for a batter doesn't change across a season. The probability p represents the hitting ability of the particular player, and so we are saying that a batter's ability stays relatively constant across games during the season. The second important assumption is that the chance of a player getting a hit doesn't depend on his performance in previous at-bats. This means that the player can't be streaky in his batting ability—the result of a particular at-bat (hit or out) doesn't depend on how he did in his recent at-bats.

One might argue that the binomial assumptions are too simplistic—you might think that a batter's ability does change over the season, and you may have heard that particular players are streaky who tend to go through stretches of good hitting and bad hitting. Also, a batter may have different abilities to hit against different pitchers and in different ballparks. But, although the assumptions seem a bit restrictive, the binomial model will be shown to give reasonable predictions of batting performance.

Let's illustrate the use of the binomial distribution to model the game-to-game hitting performance of two players, Aramis Ramirez and Jeff Kent. Ramirez in the 2001 season got 181 hits in 603 official at-bats for a batting average of $181/603 = .300$. Kent also had 181 hits, but in 607 at-bats for a $181/607 = .298$ batting average.

Ramirez played in 158 games during the 2001 season. Table 6.1 categorizes these games by the number of at-bats (AB) and the number of hits (H).

TABLE 6.1

Categorization of the 2001 game batting results of Aramis Ramirez by the number of at-bats and the number of hits.

		Hits					
		0	1	2	3	4	Total
	1	1	0	0	0	0	1
	2	4	1	1	0	0	6
Game	3	11	18	4	2	0	35
AB	4	24	50	20	2	1	97
	5	0	4	6	4	3	17
	6	0	0	1	0	1	2

Let's focus on the games where Ramirez had exactly four at-bats. We notice a lot of variation in the number of game hits. Twenty-four games Ramirez was hitless, 50 games Ramirez was 1 for 4, 20 games he was 2 for 4, two games he was 3 for 4, and one game, he was 4 for 4. Can this variation in the number of hits be explained by the binomial distribution?

To find probabilities using the binomial formula, we need to specify n and p. Since we're only considering games where Ramirez has four (official) opportunities to bat, $n = 4$, a reasonable guess at p is .3, the batting average of Ramirez for the entire 2001 season. Here X denotes the number of hits for Ramirez for this four-AB game.

Using these values of n and p, we compute the probabilities for the five possible values of X in Table 6.2. Also, since Ramirez has 97 of these games, we can also compute the expected number of games where he has different numbers of hits by multiplying these probabilities by 97.

TABLE 6.2

Probability and expected number of games (in 97 games) for Aramis Ramirez to have different hit numbers using the binomial formula with $n = 4$ and $p = .3$.

X	Binomial probability	Expected number	Observed
0	$\binom{4}{0}.3^0(1-.3)^{4-0} = 0.2401$	$97 \times 0.2401 = 23.3$	24
1	$\binom{4}{1}.3^1(1-.3)^{4-1} = 0.4116$	$97 \times 0.4116 = 39.9$	50
2	$\binom{4}{2}.3^2(1-.3)^{4-2} = 0.2646$	$97 \times 0.2646 = 25.7$	20
3	$\binom{4}{3}.3^3(1-.3)^{4-3} = 0.0756$	$97 \times 0.0756 = 7.3$	2
4	$\binom{4}{4}.3^4(1-.3)^{4-4} = 0.0081$	$97 \times 0.0081 = 0.8$	1

To see if the binomial distribution is a good fit to Ramirez's hitting data, we compare the expected counts using the binomial formula with the actual observed counts in the table. Comparing the two columns, we see that:

- Ramirez had more 1-hit games than we would expect.

- Ramirez had fewer 2 and 3-hit games than we expect.

These observations don't necessarily mean that the binomial formula is a poor fit to Ramirez's data. For example, we expect 50 heads in 100 tosses of a fair coin, but the probability of getting exactly 50 heads is very small. The question is if the differences between Ramirez's observed numbers and the expected numbers can be explained by chance, or if the differences really reflect a misfit of the binomial model.

To answer this question, we perform a simple simulation. We assume that Ramirez's hitting probability is $p = .3$ and we simulate many sequences of 97 four-AB games using the binomial model with $n = 4$ and $p = .3$. Each time, we simulate the season data, we keep track of the number of no-hit games, one-hit games, and so on. In Figure 6.1, we graph the numbers of game hits for the 100 simulated seasons using dots, and graph Ramirez's numbers by a solid line. Comparing Ramirez's with the dotplots, we see that:

- Ramirez's number (50) of one-hit games seems unusually high—for only two of the 100 simulated seasons were there 50 or more one-hit games.

- Ramirez's number of three-hit games (2) seems a bit low. In the 100 simulated seasons, the hitters got on the average eight three-hit games and it was rare to see two or fewer three-hit games.

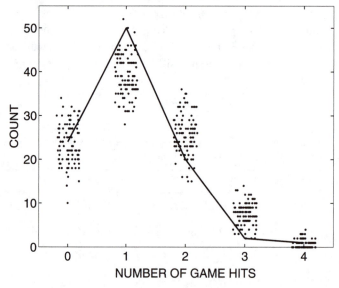

FIGURE 6.1

Graph of count of games (out of 97) with different number of game hits from the simulated binomial distribution with $n = 4$ and $p = .3$. The observed number of 0-hit, 1-hit, 2-hit, 3-hit, and 4-hit games for Aramis Ramirez is displayed by a solid line.

So for this particular player, the binomial formula doesn't seem to fit the data that well. Here is one possible explanation for Ramirez's shortage of 2- and 3-hit games. Ramirez

TABLE 6.3

Categorization of the 2001 game batting results for Jeff Kent by number of at-bats and number of hits.

		Hits					
		0	1	2	3	4	Total
	1	3	1	0	0	0	4
	2	7	3	1	0	0	11
Game	3	15	19	4	0	0	38
AB	4	16	29	16	5	1	67
	5	4	11	13	7	1	36
	6	0	1	1	0	1	2
	7	0	0	1	0	0	1

may get lots of hits when the other team's pitching is poor. In this case, Ramirez's team-mates will also get lots of hits, and it likely that he will get more than four at-bats. In other words, the number of at-bats is informative about the number of hits he will get.

Let's try Jeff Kent, who had an approximate .300 batting average in the 2001 season. Table 6.3 shows the number of hits and at-bats for Kent in the 2001 season. We again focus on the games where Kent had exactly four at-bats and see if the numbers of hits in these games match well to a binomial formula.

As above, we compute the probabilities and expected counts in Table 6.4, assuming a binomial distribution with $n = 4$ and $p = .3$.

TABLE 6.4

Probability and expected number of games (in 67 games) for Jeff Kent to have different hit numbers using the binomial formula with $n = 4$ and $p = .3$.

X	Binomial probability	Expected number	Observed
0	0.2401	$67 \times 0.2401 = 16.1$	16
1	0.4116	$67 \times 0.4116 = 27.6$	29
2	0.2646	$67 \times 0.2646 = 17.7$	16
3	0.0756	$67 \times 0.0756 = 5.1$	5
4	0.0081	$67 \times 0.0081 = 0.5$	1

Here we see that Kent's observed counts are very close to the expected counts assuming the binomial formula. To confirm this general statement, we perform the same simulation as we did above. We simulate a season of 67 four-AB games and record the number of hits in each game. This process is repeated a total of 100 seasons and the dots in Figure 6.2 show the number of hits the simulated player got in these games. Here Kent's numbers (the solid line) appear to fall in the middle of the dot distributions for all five possible hit numbers. A binomial distribution seems to mimic Jeff Kent's game-to-game hitting variation very well.

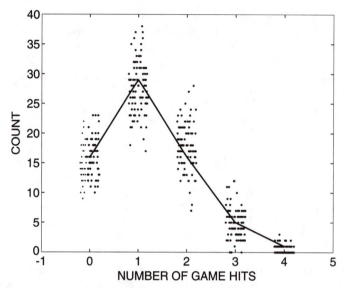

FIGURE 6.2

Graph of count of games (out of 67) with different number of game hits from the simulated binomial distribution with $n = 4$ and $p = .3$. The observed number of 0-hit, 1-hit, 2-hit, 3-hit, and 4-hit games for Jeff Kent is displayed by a solid line.

What if we looked at a large number of batters and tried to fit a binomial distribution to the daily hit numbers for each player? What would we find? In my experience, the binomial probability distribution tends to be a pretty good fit to hitting data. Although it might be insulting to say this to a baseball hitter, his variation of hitting performance across games resembles the same type of chance variation you get from tossing a coin many times where the probability of heads matches the batter's true average.

Case Study 6-2: Modeling Runs Scored: Getting on Base

Topics Covered: Negative binomial distribution, expected counts, Pearson residuals   In the next two case studies, we discuss the use of probability distributions to model the number of runs that a team scores.

How does a team score runs? It is a two-step process. First batters need to get on-base by means of hits, walks, errors, or hit-by-pitches. Second, once players get on-base, runs are generally scored by hits by other batters that move the base runners to home plate. Hitters are valuable if they are successful in getting on-base, or if they are effective in driving home runners. One measure of the ability of a batter to get on base is the on-base percentage (OBP). The ability to drive runners home is typically measured by the runs batted in or RBI.

Let R denote the number of runs scored by a particular team during a half-inning of a baseball game. We are interested in modeling the variation in the variable R by means of a probability distribution. We will do this in two steps:

1. We first construct a model for B, the number of players that come to bat during a half-inning.

2. We then find a suitable model for the runs scored (R) given that we know how many batters came to bat in the inning.

Here we focus on the number of hitters (B) that come to bat. The random variation in B can be modeled by a well-known probability distribution for experiments consisting of a sequence of "yes" or "no" outcomes (so-called Bernoulli trials).

During a half-inning, a number of players come to bat. Each player will either

OB: get on-base without creating an out

or

OUT: create an out.

Assume that the probability that any player creates an out is equal to p and the outcomes of different players in the inning are independent. Then the results of the batters in the inning can be regarded as independent Bernoulli trials with probability of creating an out, p. Players will continue to come to bat until the number of outs, or the number of successes, is equal to 3. If the batter results are independent Bernoulli trials, then the number of trials until the 3rd success, B, is distributed according to a negative binomial distribution where

$$\Pr(B = b) = \binom{b-1}{2} p^3 (1-p)^{b-3}, \quad b = 3, 4, \ldots.$$

Let's apply this formula to model the number of batters per inning for the 1987 Chicago Cubs. To compute this negative binomial formula, we need only estimate p, the chance that a batter creates an out. For the 1987 Cubs, the team on-base percentage is

$$\text{OBP} = .325$$

and so the probability that a player creates an out can be estimated by

$$p = 1 - .325 = .675.$$

In Table 6.5, we show a table of these negative binomial probabilities. In the 1987 baseball season, the Cubs batted in 1443 innings. We can obtain the expected number of innings with three batters, with four batters, and so on by multiplying the negative binomial probabilities by 1443.

TABLE 6.5
Probability of having different numbers of batters in a half-inning from a negative binomial distribution where b is the number of batters until the third out with $p = .675$.

b	3	4	5	6	7	8	9	10 or more
Probability	0.3075	0.2999	0.1949	0.1056	0.0515	0.0234	0.0101	0.0071
Expected number	443.8	432.7	281.3	152.3	74.3	33.8	14.6	10.2

TABLE 6.6
Probability of having different numbers of batters, the observed numbers from the 1987 Cubs season, and the Pearson residuals comparing the observed and expected counts.

B	3	4	5	6	7	8	9	10 or more
Probability	0.3075	0.2999	0.1949	0.1056	0.0515	0.0234	0.0101	0.0071
Expected number (e)	443.8	432.7	281.3	152.3	74.3	33.8	14.6	10.2
Observed (o)	512	426	269	127	62	25	15	7
$r = \dfrac{(o-e)^2}{e}$	10.48	0.10	0.54	4.20	2.04	2.29	0.01	1.00

Are these probabilities and expected counts a reasonable match to the actual on-base production of the Cubs during the 1987 season? To check, Table 6.6 also includes the observed number of innings where the Cubs had different numbers of batters. A standard way of gauging the difference between the observed and expected counts is by means of the Pearson residual

$$r = \frac{(\text{observed count} - \text{expected count})^2}{\text{expected count}}.$$

When the observed count is far from its corresponding expected count, the residual will be large. A value of the residual that is 4 or larger indicates a significant discrepancy between the observed count and the fitted count assuming the negative binomial model. Looking at Table 6.6, we see the only large residual is the one that corresponds to $B = 3$. We would expect (from the model) that the Cubs would have three batters (that is, a one-two-three inning) for 443 innings when in actuality they had three batters for 512 innings, 69 more. Otherwise, the negative binomial distribution appears to be a reasonable match to the Cubs on-base data.

Can we offer any explanation for the lack of fit of our model? We are assuming that the probability that a player gets on-base is constant for all players. We know that players are not equally proficient in getting on-base and the batters at the bottom of the order are weak hitters with relatively small values of p. So the Cubs' large number of three-batter innings may be a reflection of the innings where the bottom of the batting order is hitting.

Case Study 6-3: Modeling Runs Scored: Advancing the Runners to Home

Topics Covered: Multinomial probability distribution, independence, expected counts, Pearson residuals   In the previous study, we focused on the number of hitters that come to bat during a particular half-inning. In this case study, we consider the problem of modeling the number of runs scored in a half-inning. The runs scored depends significantly on the number of players B that bat during an inning. So we focus on modeling, with a simple probability distribution, the number of runs scored conditional on the fact that exactly B batters come to bat.

If six hitters come to bat in an inning, we know that three hitters created outs (there are three outs in an inning), and so the remaining $6 - 3 = 3$ hitters get on-base. How many runs can the team score when three runners get on-base? Well, the team could leave the bases loaded and score no runs. Or maybe all of the runners will score, resulting in three runs. Actually, the number of runs scored can be 0, 1, 2, or 3.

The exact number of runs scored depends heavily on the type of hit or non-hit of the batters that reach base. There are five possibilities for this on-base event:

On-base event	Abbreviation
walk or hit-by-pitch	W
single	1
double	2
triple	3
home run	H

(We place a walk and a hit-by-pitch in the same classification, since both events have the same effect on runners on-base.) Each team will tend to hit different proportions for these five events. Some teams will rely on power and hit a high fraction of doubles and home runs; other teams may rely on their ability to draw a walk and have a high proportion of "walk/hit-by-pitch" events. We let f_0, f_1, f_2, f_3, f_4 denote the probabilities that the on-base event of the team is a walk/hit-by-pitch, single, double, triple, and home run, respectively. We call the proportions $(f_0, f_1, f_2, f_3, f_4)$ the *on-base profile* of the team.

Suppose that you are given a particular sequence of on-base events. For example, suppose you are told that the first batter on-base gets a double, the next one singles, the next one draws a walk, and the last on-base person singles. (We abbreviate this sequence as "21W1.") Then by making some assumptions about runner advancement, we can figure out how many runs will score.

We will use the following runner advancement assumptions based on what is typical in a baseball game.

1. A single will move a runner from first base to third base, and score a runner from second or third base.

2. A double or a triple will score all runners from first, second, and third bases.

3. A home run will score all runners.

4. An out does not advance a runner and does not eliminate a runner. (That is, all outs are treated as strikeouts.)

Using these assumptions, we can compute the runs scored for any sequence of on-base events. To illustrate, Table 6.7 shows the base situation after every event in the sequence "21W1." For this particular sequence, a total of two runs were scored in the inning.

Suppose that each on-base event can be one of the five possibilities with probabilities f_0, f_1, f_2, f_3, f_4, and the on-base events for different hitters are independent. Then we can compute the probability of any sequence of events by simply multiplying the corresponding

TABLE 6.7
Illustration of runs scored in the play sequence "21W1" using our advancement assumptions.

Event	Description	Bases	Runs Scored
"2"	Double		0
"1"	Single		1
"W"	Walk		0
"1"	Single		1

probabilities. For example, the probability of the sequence "21W1" will be

$$\text{Prob}(21W1) = \text{Prob}(2) \times \text{Prob}(1) \times \text{Prob}(W) \times \text{Prob}(1)$$

$$= f_2 \times f_1 \times f_0 \times f_1.$$

Let's return to our original question—suppose a team has three on-base events. Note that at most three runs can score from three on-base sequences, since a runner can only score if he gets on base. What is the probability that the team scores 0, 1, 2 or 3 runs? We compute this by

- finding all three on-base event sequences that result in the particular number of runs scored,

- finding the probability of each on-base sequence by multiplying the f probabilities as we did above,

- adding the probabilities of all of the on-base sequences.

We illustrate this computation for one case. What is the probability of scoring exactly one run if you have three on-base events? It turns out there are 24 sequences of three on-base events that result in one run scored.

ww1	12w	23w	hww
w11	13w	3w1	hw1
w2w	2w1	31w	h1w
w3w	21w	311	h11
1w1	211	32w	h2w
111	22w	33w	h3w

So, the probability of scoring one run (given three people on-base) is

$$\text{Prob(1 run given 3 on-base events)} = f_0 f_0 f_1 + f_1 f_2 f_0 + f_2 f_3 f_0 + f_4 f_0 f_0$$
$$+ f_0 f_1 f_1 + f_1 f_3 f_0 + f_3 f_0 f_1 + f_4 f_0 f_1$$
$$+ f_0 f_2 f_0 + f_2 f_0 f_1 + f_3 f_1 f_0 + f_4 f_1 f_0$$
$$+ f_0 f_3 f_0 + f_2 f_1 f_0 + f_3 f_1 f_1 + f_4 f_1 f_1$$
$$+ f_1 f_0 f_1 + f_2 f_1 f_1 + f_3 f_2 f_0 + f_4 f_2 f_0$$
$$+ f_1 f_1 f_1 + f_2 f_2 f_0 + f_3 f_3 f_0 + f_4 f_3 f_0.$$

Using this method, we can compute the probability of scoring any number of runs given that we know how many batters get on-base.

Does this probability model explain the variation in the runs scored in an inning in baseball? To check, we again look at the 1987 Chicago Cubs. To compute the probabilities of runs scored (given a particular number of runners on base), we need only to estimate the on-base profile for the Cubs shown in Table 6.8.

TABLE 6.8
On-base profile of the 1987 Chicago Cubs.

Event	Singles	Doubles	Triples	Home runs	Walks
Count	989	244	33	209	504
Proportion	.4997	.1233	.0167	.1056	.2547

We see that, of the on-base events of the Cubs, about half (49.97%) were singles, 12% were doubles, 2% were triples, 11% were home runs, and 25% were walks. We can use this on-base profile to compute the chance that the Cubs will score any number of runs given a particular number of base runners. Actually, it is a bit tedious to compute these probabilities using formulas—it is much simpler to simulate this process and use the simulated output to compute the probabilities.

We consider four scenarios—one, two, three, and four runners on-base. Table 6.9 shows, in each case, the computation of

- the probability of scoring different numbers of runs from the model,

- the expected counts found by multiplying the probabilities by the number of times the Cubs had that number of batters on base,

- the observed counts of run numbers from the 1987 Cubs season,

- the Pearson residual

$$r = \frac{(o - e)^2}{e}$$

that compares the expected (e) and observed (o) numbers.

How do the expected counts from our model compare with the observed counts? Generally the model works very well, especially when there are three or four runners on base in the inning. Looking at the largest Pearson residuals, it seems that the model is not working

TABLE 6.9
Probability of scoring different numbers of runs from the model, the observed count of run numbers from the 1987 Cubs season, and the Pearson residuals. These quantities are given for each of four runner on-base scenarios.

One runner on-base		
	Runs Scored	
	0	1
Probability	0.899	0.101
Expected count	382.1	42.9
Observed	359	66
Residual	1.40	12.44

Two runners on-base			
	Runs Scored		
	0	1	2
Probability	0.611	0.286	0.103
Expected count	164.4	76.9	27.7
Observed	145	92	32
Residual	2.29	2.96	.67

Three runners on-base				
	Runs Scored			
	0	1	2	3
Probability	0.163	0.450	0.290	0.097
Expected count	20.7	57.2	36.8	12.3
Observed	20	55	41	11
Residual	.02	.08	.48	.14

Four runners on-base				
	Runs Scored			
	1	2	3	4
Probability	0.149	0.480	0.279	0.092
Expected count	9.2	29.8	17.3	5.7
Observed	10	32	14	6
Residual	.07	.16	.63	.02

well in two situations: scoring one run when there is one runner ($r = 12.44$), and scoring one run when there are two runners in the inning ($r = 2.96$). The Cubs actually scored one run 66 times where they had only one runner in the inning and we would predict 43 from the model. Also, when the Cubs had two runners, they scored a run 92 times compared to 77 predicted from the model.

Why doesn't the model fit well in these two situations? Remember that the run probabilities are based on our model, which assumed that runner advancement was a function only of a team's on-base profile of walks and hits. There is no allowance for base stealing, sacrifice hits, or errors that are also helpful in advancing runners. Strategies like base stealing and sacrificing are often used in baseball when the game is close and the team is trying to score a single run. These other types of run advancement may explain why a team is more likely to score a run than what we predict.

But the model produces estimates that are generally close to the actual numbers of runs scored. This is a bit surprising since we are assuming that

- each player on the team has the same on-base profile (this clearly is not true),

- on-base outcomes by different players in an inning are independent,

- as we said above, the only way to advance a runner is by means of a hit or a walk.

We could change our model to make it more realistic. (Certainly people who construct baseball simulation games such as the ones discussed in Chapter 5 would want to design a model to make it as realistic as possible.) But the above model is attractive in that it is fairly simple and helps us understand the importance of on-base events toward the goal of scoring runs.

Exercises

Leadoff Exercise. Here we use the Rickey Henderson spinner that we constructed for the leadoff exercise of Chapter 5.

(a) Suppose that Rickey plays 50 games and each game he has five plate appearances. Spin the Rickey spinner a total of 250 times, keeping track of the number of times he gets on-base for each game. Record your counts in the following table.

Number of times on-base	0	1	2	3	4	5
Count						

(b) Find the probabilities that Rickey gets on-base 0, 1, 2, 3, 4, and 5 times, if the number of times on-base has a binomial distribution with 5 trials and probability of success p. (Estimate p by the fraction of times he gets on-base for the 1990 season.)

(c) Find the expected number of games where Rickey gets on base $0, 1, \ldots, 5$ times. Compare your expected counts with the simulated counts using the spinner.

6.1. In Chapter 5, we looked at the game-to-game home run hitting of Barry Bonds during the 2001 baseball season. Suppose we focus on the games in which Bonds had at least three plate appearances. Table 6.10 gives a frequency distribution of the number of home runs that Bonds hit in these games.

TABLE 6.10
Frequency distribution of the number of home runs hit by Barry Bonds in 2001 in all games with at least three plate appearances.

Number of home runs hit	0	1	2	3
Count	87	50	8	2

    Overall in these games he hit 72 home runs—an average of .4898 home runs per game.

(a) If x denotes the number of home runs hit per game, estimate the probability that x is equal to 0, 1, 2, 3 using a Poisson density with mean equal to .4898. A *Poisson density* with mean L has probabilities given by

$$f(x) = \frac{e^{-L}L^x}{x!}, \quad x = 0, 1, 2, \ldots.$$

Put your probabilities in the table below.

Number of home runs hit x	0	1	2	3
Count	87	50	8	2
Probability				
Expected count				

(b) Find the expected number of games that Bonds would hit 0, 1, 2, 3 home runs, and put these expected counts in the same table.

(c) Use Pearson residuals to compare the observed and expected counts. Is the Poisson distribution a good fit to these data?

6.2. (Exercise 6.1 continued)   As an alternative to the Poisson distribution, perhaps the distribution of x can be modeled using a binomial distribution with $n = 4$ and $p = .1099$. Here $n = 4$ represents a typical number of opportunities for Bonds to bat each game and p is the probability that he will hit a home run in a single plate appearance. Recall from Case Study 4-1 that Bonds' rate of home run hitting in 2001 was .1099. Find the binomial probabilities, expected counts, and Pearson residuals using this binomial fit. Comment on the suitability of the fit and contrast the fit with the Poisson fit in Exercise 6.1.

6.3. A baseball team typically experiences a number of winning and losing streaks during a season. Should we be surprised by these streaks? The 162 games played by the 2001 World Champion Arizona Diamondbacks were divided into 54 groups of three consecutive games. In each group of three games, the number of wins w was recorded. A frequency distribution for w is shown in Table 6.11.

TABLE 6.11
Frequency distribution of games won in groups of three consecutive games by the Arizona Diamondbacks in 2001.

w	0	1	2	3
Count	5	16	23	10

(a) Find the probabilities that Arizona wins 0, 1, 2, and 3 games in a three-game period if w is given a binomial distribution with $n = 3$ and $p = .5679$. (In the 2001 season, the Diamondbacks had a 92–70 record for a winning fraction of $92/162 = .5679$.)

(b) Find the expected number of groups that Arizona wins 0, 1, 2, and 3 in a 162-game season assuming the binomial model.

(c) Compare the observed and expected counts by the computation of Pearson residuals. Does a binomial fit seem reasonable for these data?

6.4. Table 6.12 gives the distribution of game hits for all 4-AB games in the 2001 baseball season for Shawn Green and Brian Jordan.

TABLE 6.12
Distribution of game hits in all 4-AB games in the 2001 season by Shawn Green (2001 AVG = .297) and Brian Jordan (2001 AVG = .295).

Number of hits	0	1	2	3	4
Count for Shawn Green	21	21	30	2	1
Count for Brian Jordan	16	35	12	6	1

As in Case Study 6-1, investigate if a binomial distribution is a reasonable model of the number of game hits for each player. In each case, use the season batting average (AVG) for the binomial probability of success.

6.5. Suppose that we are interested in the random variable X that is the number of the half-inning in which the first run is scored in a baseball game. In the game

				R	H	E	LOB
Texas	040	210	000	7	8	0	7
Anaheim	012	001	001	5	15	0	10

the first run scored was in the top of the 2nd inning, so $X = 3$. In the game

				R	H	E	LOB
Milwaukee	001	100	010	3	9	1	9
Houston	206	000	30x	11	11	0	5

the first run scored was in the bottom of the 1st, so $X = 2$. We recorded X for a sample of 150 games in the 2001 season and the observed frequency distribution of X is shown in Table 6.13.

TABLE 6.13
Frequency distribution of the number of the half-inning until the first run scored for 150 games in the 2001 season.

X	1	2	3	4	5	6	7	8	9	10	11	19	Total
Count	39	39	22	14	9	4	9	5	4	2	2	1	150

We can model X by a geometric distribution, where p is the probability that a team scores in a single inning. Here X represents the number of trials (half-innings) until the first success (half-inning with a run scored). In a sample of 2400 half-innings from the 2001 baseball season, the offensive team scored in 703 of the half-innings, so the probability p can be estimated by $703/2400 = .297$.

(a) Find the probabilities that X takes on the values $1, 2, 3, \ldots$ using the geometric probability distribution with formula

$$\Pr(X = x) = p\,(1 - p)^{x-1}, \quad x = 1, 2, \ldots$$

where $p = .297$.

(b) Find the expected number of games (out of 150) in which $X = 1, 2, 3, \ldots, 19$.

(c) Compare the observed and expected counts. Is the geometric distribution a good fit to these data?

6.6. Suppose a team has the on-base profile $(f_0, f_1, f_2, f_3, f_4)$, where the fractions f_0, f_1, f_2, f_3, f_4 represent the probabilities of the events "walk" (w), "single" (1), "double" (2), "triple" (3), and "home run" (h), respectively.

(a) Suppose that the team has three on-base events during a particular half-inning. Find the probability of not scoring any runs that half-inning. (*Hint:* Using the runner advancement rules described in this chapter, the sequences of events that will not produce a run are {www, w1w, 1ww, 11w, 2ww, 3ww}.)

(b) Suppose that you have 2 on-base events during a half-inning. Find the probability of scoring 0, 1, and 2 runs in that inning. (*Hint:* There are 25 possible sequences of two on-base events—six of these sequences will not score a run and five of these sequences will score exactly two runs.)

6.7. Table 6.14 gives the observed distribution of inning runs scored for 150 games in the 2001 season (the first through eighth innings).

TABLE 6.14
Frequency distribution of the number of runs scored in an inning for 150 games in the 2001 season.

Runs scored	0	1	2	3	4	5	6	7	8	9 or more	Total
Count	1687	372	179	90	40	21	5	1	3	2	2400
Expected count											

Suppose that we apply the run scoring model described in the case studies to these data. For the 2001 teams, a typical on-base fraction is .3315, so we can estimate the probability of an out to be $p = 1 - .3315 = .6895$. Also for the 2001 teams, the on-base fractions (of a walk, single, double, triple, and home run) are given by

$$f_0 = .2649, \ f_1 = .4805, \ f_2 = .1477, \ f_3 = .0155, \ f_4 = .0914$$

By simulation, we used this model to estimate the probability that a team scores $0, 1, 2, \ldots$ runs in the half-inning. The probabilities from the model are displayed in Table 6.15.

TABLE 6.15
Probability distribution of the number of inning runs scored for using the run-scoring model.

Runs scored	0	1	2	3	4	5	6	7	8	9 or more
Probability	.697	.149	.080	.043	.018	.009	.003	.002	.000	.000

Use these probabilities to find the expected number of half-innings (out of 2400) where $0, 1, 2, \ldots$ runs would score. Put the expected counts in the first table. Comparing the observed and expected counts by the use of Pearson residuals, is the model effective in predicting the number of inning runs scored in major league games?

6.8. (Exercise 6.7 continued) One assumption in our run scoring model is that the run production ability of the team doesn't change across innings. Are there particular innings during a game where a team is more or less likely to score runs? For our sample of 150 games during the 2001 season, the mean runs scored for each inning (first through eighth) was computed. A graph of the mean runs scored is shown in Figure 6.3.

This figure illustrates that teams are indeed more or less likely to score in particular innings. Describe the variation that you see in the plot and explain why some innings are particularly good (or bad) for run production.

6.9. In the 2001 baseball season, the Colorado Rockies were very effective in scoring runs and the Tampa Bay Devil Rays were relatively ineffective in scoring. Here are some hitting statistics for the two teams.

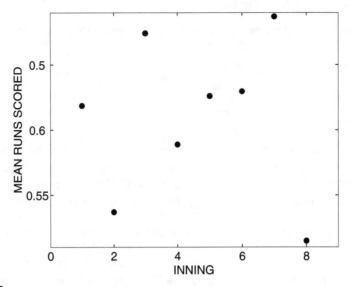

FIGURE 6.3

Plot of the mean number of runs scored in different innings for a sample of 150 games during the 2001 season.

Tm	H	2B	3B	HR	BB	OBP
COL	1663	324	61	213	511	.354
TBD	1426	311	21	121	456	.320

(a) Compare the on-base profile of the two teams.

(b) Compute the probability that the team scores at least one run with two base runners for the two teams.

(c) Compute the average number of runners per inning for the two teams.

(d) If you were the General Manager of Tampa Bay, what types of hitters would you try to sign for the 2002 season? Explain.

Further Reading

Basic discrete probability distributions, such as the binomial, negative binomial, and Poisson, are presented in Scheaffer (1995). Mosteller (1952) uses probability modeling to understand the World Series baseball competition. D'Esopo and Lefkowitz (1977), Cover and Keilers (1977), and Albert and Bennett (2001), Chapter 8, discuss probabilistic models for run production.

7

Introduction to Statistical Inference

What's On-Deck?

This chapter describes some fundamental notions about statistical inference in the context of baseball. One important idea in inference discussed in the introduction is the distinction between a player's ability and his performance. We are interested in a player's hitting ability, which can be measured by a probability p that represents the player's chance of getting on base in a single plate appearance. We don't know a player's ability p, but we learn about this value when we see the player perform in a series of games. In Case Study 7-2, we first consider the situation where you know a player's on-base probability p, and we look for basic patterns in his hitting performance in ten plate appearances. In Case Study 7-3, we use a simple simulation to describe how we can learn about a player's batting ability when we observe his performance in ten at-bats. We initially suppose that the player's on-base ability p is equally likely to be .2, .3, or .4. In the simulation, we choose a player's ability at random, and then simulate the process of having this player have ten plate appearances. We categorize the simulated players' abilities and performances in a two-way table, and we perform inference by looking at the abilities of the players corresponding to a given number of on-base events in ten plate appearances. Case Study 7-4 illustrates the use of a basic formula for an interval estimate for a batter's ability p of a particular probability content. We use this interval formula in Case Study 7-5 to compare the hitting abilities of two great hitters, Wade Boggs and Tony Gwynn. We will see that it is difficult to distinguish the hitting abilities of two players based on only one season of data, but we can make a clearer distinction by looking at the pattern of hitting of the two players over all of the seasons of their careers.

Case Study 7-1: Ability and Performance

Topics Covered: Distinction between ability and performance   Recently a statistics class of 28 students played a spinner baseball game between Hall of Fame players of the American League and the National League. In this game, the AL defeated the NL 10-0. Why? I gave the students four possible explanations in a test question:

1. The AL was lucky—the spins went the right way.
2. The AL players were better than the NL players.
3. The win was a result of luck *and* the fact that the AL players were better.
4. There was some cheating going on.

How did the students answer this question? Here's the tally:

Answer	Tally
Luck	12
Ability	1
Luck & Ability	14
Cheating	1

Practically all of the students thought that luck played a role in the AL win.

Now what if I asked the students the same question regarding the Yankees' World Series win against the Mets in 2000? Would they also say that the Yankees' win was partly due to luck? I'm guessing that most of the students would say that the Yankees won because they were more skilled than the Mets. But actually luck or chance variation plays a big role in baseball games.

Why did the students think differently about the spinner game and the real World Series? Well, it is obvious that chance plays a big role in the spinner game since all of the outcomes depend on the spins of the spinners. We don't see any obvious dice or spinners in baseball games. But, if you think about it, there are a lot of chance elements in a baseball game (how the ball moves through the infield grass, how the bat hits the ball, how a player fields a ground ball, etc.) that have random or unpredictable elements and this randomness can affect the outcome of the game.

> One primary role of the statistician is to understand how much of the variation in baseball data (and other data as well) can be explained by chance and how much is due to some "real" cause, like the skill of a player.

Baseball Hitting—ability and performance. It's helpful to look at the dictionary's definition of these two words. **Ability** is (1) the power to do or act, (2) skill, (3) power to do some special thing, natural gift, talent. **Performance** is (1) the act of carrying out, doing; performing, (2) a thing performed; act; deed.

When we say a player has great ability to hit, we are talking about his skill or his gift to hit the ball. This player might have a great eye for the ball, have a nice swing, and make good contact with the ball. He may be muscular with the capability to hit the ball a long way. In contrast, the performance of a player is actually how he hits during games. Here a batter's performance is simply the record of hitting that you observe in the box scores. If Barry Bonds hit two home runs today, he had a great batting performance.

These two words are connected. If a player has great batting ability, he will generally exhibit great batting performances. But it is important to distinguish between ability and performance. Barry Bonds may be 1 for 10 in two playoff games. Although he had a weak batting performance, it doesn't mean that he's turned into a bad hitter. Likewise, a hitter who is 4-4 for one game isn't necessarily a much better hitter. He may actually be a mediocre hitter and happened to be lucky or fortunate that particular game.

We say that Mickey Mantle was a great hitter. Most baseball fans would agree that Mantle had great batting ability. Why? Because he had one great season? No. He had a number of great seasons. In other words, he exhibited a pattern of great batting behavior. Roger Maris, in contrast, had only a few great hitting seasons (especially 1961 when he hit 61 home runs). Since Maris didn't show a consistent behavior of great hitting for many seasons, there is some doubt that he really had great batting ability. In fact, some baseball people think that he was a bit lucky in 1961—he just happened to be the right hitter at the right time.

In statistical inference, the objective is to learn about a player's ability based on his performance in the field. We will see that it is difficult to learn a lot about a player's ability based on the performance of the player in a single season. In the next case study, we'll look at the relationship between performance and ability more carefully.

Case Study 7-2: Simulating a Batter's Performance if His Ability Is Known

Topics Covered: Random spinner as a probability model, simulation of a binomial experiment using dice   We represent ability by means of a **probability model**. This is a simple randomization device with known properties. This model is said to be *realistic* if it generates baseball data similar to what is actually observed. A basic probability model is one that we have already seen—a spinner. Figure 7.1 displays a spinner that represents the outcome of a plate appearance.

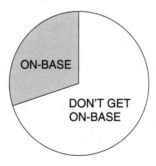

FIGURE 7.1
Random spinner to represent the outcome of a plate appearance.

In this spinner there are two possible outcomes—on-base and not on-base. The ratio of the area of the on-base region to the area of the entire circle in the spinner represents the probability that the player will get on-base.

Another convenient probability model is based on a die. Imagine that you have a die with ten sides, labeled $0, 1, \ldots, 9$. Each side has the same chance of being rolled and each side has probability $\frac{1}{10}$. Then we can represent the result of a plate appearance by a roll of this die. Suppose that we decide to let

- rolls 1, 2, 3, 4 correspond to on-base,

- rolls 0, 5, 6, 7, 8, 9 correspond to not on-base.

The 10-sided die represents the player's ability to get on-base. A roll of this 10-sided die represents the performance of the player on a single at bat.

Let's say we specify a player's true ability. For example, for Roberto Alomar, we specify a probability of .400 of getting on base. What kind of performance can we expect, say in ten plate appearances?

In a statistics class, 10-sided dice were handed out to the students and a number of simulations were performed. In each simulation, we rolled the die ten times, representing Alomar's ten PAs. Remember an on-base event corresponds to a 1, 2, 3, 4 on the die, and we keep track of the number of rolls that were either 1, 2, 3, 4. Each of the 30 students performed five simulations—the results of the 150 simulations are shown in Table 7.1. This table represents the performance of the batter over 150 10-plate appearance periods.

TABLE 7.1
Frequency distribution of number of times on-base using dice with 150 simulations.

Number of times on-base	0	1	2	3	4	5	6	7	8	9	10
Count	1	7	18	40	31	32	11	8	2	0	0
Probability	.007	.047	.120	.267	.207	.213	.073	.053	.013	0	0

After we do this a sufficient number of times, we answer the following questions:

- What was the most likely number of times on-base for Alomar? Here the most likely outcome in our 150 simulations was 3 times on-base—this occurred 40 times.

- What is the chance of this most likely number? The estimated probability of three times on-base is $40/150 = .267$.

- What is the probability that Alomar will get on-base two or more times? In our 150 simulations, we see that Alomar got on-base once or never 8 times. So, by subtraction, the frequency of "2 or more times" is $150 - 8 = 142$. The estimated probability of "2 or more times" is $142/150 = .947$.

- What is the probability that Alomar will not get on-base during these ten PAs? In our 150 simulations, Alomar didn't get on-base one time. So the estimated probability of "not getting on-base" is $1/150 = .007$.

In statistical inference, we're actually interested in looking at ability and performance the opposite way. We observe a hitter's performance. What does that tell us about a player's ability? This is what we'll talk about in the next section.

Case Study 7-3: Learning about a Batter's Ability

Topics Covered: Modeling ability by using a spinner with probability p, simulating hitting data for a given ability, simulating abilities and performance, Bayes' rule, finding the most likely ability for a given performance   We represent a player's ability by means of a probability model. Imagine that a player's talent to get on-base is represented by the spinner displayed in Figure 7.1. For a circle of area 1, the area of the on-base region is equal to the player's ability to get on-base. We call this area p—is our measure of a player's hitting ability.

Suppose we know a player's hitting ability. Specifically, suppose we know his on-base probability p is equal to .4. This player comes to bat ten times during a doubleheader. How many times will he get on-base?

We did this simulation in a statistics class in the previous case study using dice. These probabilities aren't too precise, since the simulation was performed only 150 times. On the computer I performed this experiment 1000 times. We assume that the player has an on-base probability of $p = .4$, and the results of 1000 doubleheaders were simulated where the player had ten opportunities to hit in each doubleheader. A frequency distribution of the times on-base is shown in Table 7.2.

TABLE 7.2
Frequency distribution of number of times on-base using a computer with 1000 simulations.

Number of times on-base	0	1	2	3	4	5	6	7	8	9	10	Total
Count	4	39	130	208	246	209	98	49	14	3	0	1000
Probability	.004	.039	.130	.208	.246	.209	.098	.049	.014	.003	0	1

Note that the most likely outcome for this hitter is four times on-base. This makes sense—if a player with probability of .4 of getting on-base has ten chances, one would expect him to get on-base $10(.4) = 4$ times.

Above we assumed that our player had an on-base probability of .4. What if his on-base probability was $p = .3$? On the computer I simulated the result of 1000 doubleheaders assuming $p = .3$. The results are shown in Table 7.3 and compared with the results when $p = .4$.

TABLE 7.3
Frequency distribution of number of times on-base using computer with 1000 simulations for the cases where $p = .4$ and where $p = .3$.

	Number of times on-base	0	1	2	3	4	5	6	7	8	9	10	Total
$p = .4$	Count		4	39	130	208	246	209	98	49	14	3	0	1000
	Probability	.004	.039	.130	.208	.246	.209	.098	.049	.014	.003	0	1
$p = .3$	Count		35	131	208	261	210	101	39	13	2	0	0	1000
	Probability	.035	.131	.208	.261	.210	.101	.039	.013	.002	0	0	1

If we compare the results when $p = .4$ with the results when $p = .3$, we see some differences. If $p = .4$, the most likely outcome is four times on-base; in the case when $p = .3$ the most likely value is three times on-base.

If a batter has a true .3 on-base probability, is it accurate to say that he's sure to get on-base three times (out of ten)? No. We see that the probability of three times on-base is only .261—it's actually more likely (.739) that he won't get on-base exactly three times.

In our simulations above, we assumed that we knew the player's ability (the value of p) and we looked at possible outcomes in a doubleheader (the batter's performance in ten plate appearances). We actually want to solve the inverse problem. If a player gets on-base four times (out of ten), what does that say about the player's ability (value of p)?

We use a simple simulation to see how we can learn about a batter's hitting ability. Suppose there is a manager named Casey who has a dugout of players who are equally divided between hitters of three abilities: the crummy hitters who have a true on-base probability $p = .2$, the mediocre hitters who have $p = .3$, and the good hitters who have $p = .4$. Suppose Casey picks a player from the dugout at random and the player gets 10 chances to hit. Casey observes

- the player's ability (value of p),

- the player's performance—the value of $x = $ number of times on-base.

Three spinners were used in this simulation—one with an on-base probability of .2, another with an on-base probability of .3, and the third with an on-base probability of .4. We first choose a spinner at random. We rolled a single die—if the die roll was 1 or 2, we used the $p = .2$ spinner; if we rolled 3 or 4, we used the $p = .3$ spinner; if the roll was 5 or 6, we used the $p = .4$ spinner. We then spin the chosen spinner ten times.

In one particular simulation, we chose the bad spinner ($p = .2$). We spun it ten times with the results

NNNNNBNNNB

(N means not on-base, B means on-base.) So we observed $p = .2$ and $x = 2$. We repeat this simulation 1000 times—we classify all of the results in Table 7.4 by ability (value of p) and performance (value of x).

Before we observed any hitting data, what is the chance that the spinner chosen is a $p = .2$ spinner? Each spinner has the same chance of being chosen, so the probability that

TABLE 7.4
Two-way table of simulation results classified by the player's ability (value of p) and his performance (value of x).

		\multicolumn{11}{c}{Performance (value of x)}									
		0	1	2	3	4	5	6	7	8	9	10
Ability (value of p)	.2	38	105	113	61	22	5	0	0	0	0	0
	.3	14	35	74	75	83	36	12	3	1	0	0
	.4	3	9	34	76	90	54	46	4	6	1	0

a $p = .2$ spinner is chosen is $\frac{1}{3}$. That is,

$$\text{Prob}(p = .2) = \text{Prob}(p = .3) = \text{Prob}(p = .4) = \frac{1}{3}.$$

Now suppose that we observe four times on-base for our player, that is, $x = 4$. In Table 7.5, we focus on the column where $x = 4$.

TABLE 7.5
Two-way table of simulation results of ability and performance where we restrict attention to the performance $x = 4$.

		\multicolumn{11}{c}{Performance (value of x)}
		0	1	2	3	**4**	5	6	7	8	9	10
Ability (value of p)	.2	38	105	113	61	**22**	5	0	0	0	0	0
	.3	14	35	74	75	**83**	36	12	3	1	0	0
	.4	3	9	34	76	**90**	54	46	4	6	1	0

We see that we observed 4 on-base a total of $22 + 83 + 90 = 195$ times. Of these 195 times, 22 corresponded to a hitting probability of $p = .2$, 83 corresponded to a probability of $p = .3$, and 90 corresponded to a hitting probability of $p = .4$. Converting these counts to probabilities

		Count	Probability
Ability (value of p)	.2	22	$22/195 = .11$
	.3	83	$83/195 = .43$
	.4	90	$90/195 = .46$

we see that this player is likely to be either a $p = .3$ hitter or a $p = .4$ hitter. The probability that he is a $p = .2$ hitter is only about 11%.

What if the hitter got on-base six times? (That is, you observed $x = 6$.) Then you would focus on the column of the table corresponding to $x = 6$.

		Count	Probability
Ability (value of p)	.2	0	$0/58 = 0$
	.3	12	$12/58 = .21$
	.4	46	$45/58 = .79$

Here it is highly likely (probability .79) that $p = .4$. So if you observe six times on-base, you can conclude that the hitter is likely a true 40% hitter. Also you are pretty sure that the hitter is not a .2 hitter, since $\text{Prob}(p = .2) = 0$.

Case Study 7-4: Interval Estimates for Ability

Topics Covered: Learning about a true ability (value of p) by means of an interval estimate, subjective interpretation of probability.

A Different Way of Thinking About a Probability. Al Gore said during the 2000 Presidential campaign that he had a 50% chance of winning the presidential election. What does 50% mean? It's a probability, but not in the relative frequency sense—the election is a one-time event and it doesn't make sense to imagine having many elections between Bush and Gore. Al Gore's 50% is a probability, but it represents Gore's *degree of belief* in winning the election. This interpretation of probability is *subjective*; different people can assign different probabilities to the event "Gore will win," since different people will have different opinions about the likelihood of the event. This interpretation of probability is relevant here. We will use probability to represent our beliefs about different batting abilities of a player.

Suppose we are interested in learning about the batting ability of Alex Rodriguez. Recall that we represent the hitting ability of a player by a spinner, where there are two events, Hit and Out, and the size of the Hit area represents the probability of a Hit. We denote this hitting probability by p—this is a player's **true batting average**.

The first step in learning about Rodriguez's hitting ability is to list some possible values of p. Let's assume that Rodriguez's hitting probability could be

$$p = .1, .2, .3, .4, .5, .6, .7, .8, .9$$

and each of these nine values are equally likely. So Prob(Rodriguez is a .100 hitter) = Prob(Rodriguez is a .200 hitter) = $\cdots$ = Prob(Rodriguez is a .900 hitter) = $\frac{1}{9}$.

Now you are probably thinking these assumptions are silly. Rodriguez can't be a .100 or .900 hitter, and even if there were nine possible batting abilities, it doesn't make sense to assign each value of p the same probability. (Rodriguez is more likely to be a .300 or .400 hitter.) You're right—these are unrealistic assumptions, but it makes the calculations to follow easy to explain.

Next, we let Rodriguez bat 20 times and we observe x = number of hits. Suppose we observe $x = 6$ (Rodriguez gets six hits out of 20 at-bats).

What can we say about Rodriguez's true batting ability p? We use the simulation scheme that we introduced in Case Study 7-3 to learn about Rodriguez's hitting ability.

1. We first choose an ability at random from the values $p = .1, .2, \ldots, .9$. Imagine nine spinners corresponding to the nine possible hitting probabilities and we choose one spinner at random.

2. We spin the chosen spinner (from part 1) 20 times and we count x = number of hits.

We repeat this process (choose a spinner and spin 20 times) a total of 10,000 times (on a computer). The results are displayed in Table 7.6.

Remember that each time we do the simulation, we select an ability p (a spinner) and observe a hit number x. Look at the first count in the table, 143, in the upper left corner. This means that 143 times we chose the $p = .1$ spinner and observed no hits ($x = 0$).

TABLE 7.6

Two-way table of counts from simulation where there are nine possible abilities $p = .1, \ldots, .9$, and the batter comes to bat 20 times and gets x hits.

		0	1	2	3	4	5	6	7	8	9	10	11	12	13	14	15	16	17	18	19	20
															x (number of hits)							
	0.1	143	276	334	207	99	45	8	4	0	0	0	0	0	0	0	0	0	0	0	0	0
	0.2	14	65	147	216	247	200	130	60	25	7	3	1	0	0	0	0	0	0	0	0	0
	0.3	0	13	23	78	167	199	195	206	110	62	40	11	3	0	0	0	0	0	0	0	0
p	0.4	0	2	6	13	32	90	133	184	195	203	138	63	4	8	8	8	1	0	0	0	0
(probability)	0.5	0	0	0	4	5	22	42	80	107	179	201	188	142	7	3	47	19	1	0	0	0
	0.6	0	0	0	0	1	1	5	17	44	82	126	167	210	201	134	92	38	17	3	0	0
	0.7	0	0	0	0	0	0	0	1	3	14	38	67	118	182	214	216	151	63	36	10	0
	0.8	0	0	0	0	0	0	0	2	0	0	3	4	22	60	110	189	231	232	155	58	8
	0.9	0	0	0	0	0	0	0	0	0	0	0	0	0	4	17	34	102	182	317	289	158

TABLE 7.7

Simulated values of ability p when the hitter gets $x = 6$ hits.

		Count	Probability
	0.1	8	0.016
	0.2	130	0.253
	0.3	195	0.380
p	0.4	133	0.259
(ability)	0.5	42	0.082
	0.6	5	0.010
	0.7	0	0.000
	0.8	0	0.000
	0.9	0	0.000
	Total	513	

Remember Rodriguez got six hits (out of 20 AB). To see what we have learned about Rodriguez's batting ability, we focus on the $x = 6$ column of the table, shown in Table 7.7.

We convert the counts to probabilities by dividing each count by the total. These probabilities represent the likelihoods of Rodriguez having different batting abilities. From the table we see

- Prob(Rodriguez has a true .200 AVG) $= 130/513 = .253$.

- Prob(Rodriguez's ability is between $p = .2$ and $.4 =$ Prob($p = .2, .3, .4$) $= .253 + .380 + .259 = .892$.

We'd like to find an interval of ability values that are very likely. We call this interval a **probability interval** for the true batting average p.

To find a probability interval, we use the following table. In the left column, we put values of p, from most likely to least likely, and the second column contains the total probability content of these ability values.

1. We note from the table below that $p = .3$ is the most likely ability for Rodriguez—we put the value (.3) in the first column and the associated probability (.380) in the second column.

2. The next most likely value of p is $p = .4$—we put this value in the first column and the probability (.259) in the second column.

3. We continue doing this until the total probability (the sum of the probabilities) in the second column is a large number, say between 80 and 95 percent.

Values of p	Probability
.3	.380
.4	.259
.2	.253
Total	.892

We see that the values {.3, .4, .2} or the interval [.2, .4] is an 89.2% probability interval for Rodriguez's ability p. The chance that Rodriguez actually has one of these abilities is 89.2%. In other words, we are 89.2% confident that Rodriguez's true batting average is between .2 and .5.

Actually this is very little information since we know nearly every player's batting average falls between .2 and .4. We really haven't learned much about ability based on only 20 AB. We need much more data to get a better handle on a player's ability.

There is a simpler recipe for constructing this interval estimate for a player's true batting average p.

1. We first compute a player's observed AVG $\hat{p} = x/\text{AB}$ (this is his reported batting average).

2. We compute the **standard error** (SE), which is a measure of the accuracy of $\hat{p}$ to estimate a true batting average p:

$$\text{SE} = \sqrt{\hat{p}(1 - \hat{p})/\text{AB}}.$$

3. A probability interval of content PROB will have the general form

$$(\hat{p} - z\,\text{SE}, \hat{p} + z\,\text{SE})$$

where z is a value of a standard normal density such that the upper tail area is equal to $(1 - \text{PROB})/2$. Table 7.8 gives values of z for several choices of PROB.

We illustrate using this recipe. Rodriguez got six hits in 20 at-bats, so

$$\hat{p} = \frac{6}{20} = .3.$$

The standard error is

$$\text{SE} = \sqrt{.3(1 - .3)/20} = .102.$$

TABLE 7.8

Value z of a standard normal density corresponding to various values of the probability content PROB.

PROB	Upper tail area	z
.8	.1	1.28
.892	.054	1.607
.9	.05	1.645
.95	.025	1.96

Above we found an 89.2% probability interval for p. From Table 7.8, we see that if we are interested in a probability content of PROB $= .892$, we should use a value of $z = 1.607$. So a 90% interval estimate for Rodriguez's batting ability p is

$$.3 - 1.607(.102) \text{ to } .3 + 1.607(.102) = [.136, .464].$$

This gives a similar answer to what we got using our simulation method.

Generally, you learn more about a player's true batting average by observing more AB. Suppose Rodriguez plays for LA next year and has a good season—210 hits in 600 AB.

His observed batting average is $\hat{p} = 210/600 = .350$. We compute

$$SE = \sqrt{(.35(1 - .35)/600} = .019.$$

A 90% probability interval for Rodriguez's p would be

$$.350 - 1.645(.019) \text{ to } .350 + 1.645(.019) = (.319, .381).$$

Rodriguez's batting ability in 2001 could be as low as .319 or as high as .381. So we don't learn as much about a player's true batting average as we might expect.

Case Study 7-5: Comparing Wade Boggs and Tony Gwynn

Topics Covered: Interval estimates for a proportion, comparing proportions by use of interval estimates, time-series plots   In this case study we compare two of the greatest hitters in recent years, Wade Boggs and Tony Gwynn. Boggs played many years for the Red Sox. He was very effective in getting his bat on the ball (a so-called contact hitter) and won many batting crowns for the best batting average. (He's also known for his preference in diet—he had chicken before every game.) Gwynn played for the Padres his entire career and retired in 2001. He also is considered a great contact hitter. Michael Schell, in *Baseball's All-Time Greatest Hitters*, rates Gwynn as the best hitter of all time with respect to batting average.

Table 7.9 presents the batting averages for Boggs and Gwynn for all years that they played in the major leagues. I give the ages for each player each season—we'll make comparisons of the players for different ages.

Let's focus on the batting averages when both players were 31. At this age, Gwynn had 168 hits in 530 at-bats for a batting average of .317; Boggs had 205 hits in 621 at-bats for an average of .330. Who was a better hitter at age 31?

TABLE 7.9

Career batting statistics for Tony Gwynn and Wade Boggs.

	Tony Gwynn				Wade Boggs		
Age	AB	H	AVG	Age	AB	H	AVG
22	190	55	0.289	24	338	118	0.349
23	304	94	0.309	25	582	210	0.361
24	606	213	0.351	26	625	203	0.325
25	622	197	0.317	27	653	240	0.368
26	642	211	0.329	28	580	207	0.357
27	589	218	0.370	29	551	200	0.363
28	521	163	0.313	30	584	214	0.366
29	604	203	0.336	31	621	205	0.330
30	573	177	0.309	32	619	187	0.302
31	530	168	0.317	33	546	181	0.332
32	520	165	0.317	34	514	133	0.259
33	489	175	0.358	35	560	169	0.302
34	419	165	0.394	36	366	125	0.342
35	535	197	0.368	37	460	149	0.324
36	451	159	0.353	38	501	156	0.311
37	592	220	0.372	39	353	103	0.292
38	461	148	0.321	40	435	122	0.280
39	411	139	0.338	41	292	88	0.301
40	127	41	0.323				
41	102	33	0.324				
Lifetime			0.338	Lifetime			.328

I didn't ask who performed better at age 31. That question is simple to answer—Boggs had the higher batting average (.330 compared to .317). I'm actually asking about the hitting abilities of Gwynn and Boggs. Can we say that Boggs had a better hitting ability that year?

Remember we represent a hitter's ability by means of a spinner where the Hit area is equal to p. So Gwynn and Boggs this year have abilities (or spinners) with hitting probabilities of p_G and p_B. (We call these the true batting averages.) Can we say that p_B is larger than p_G?

We answer this question by computing probability intervals for the two true batting averages. We use 95% intervals—recall the formula from Case Study 7-4. So for Boggs, we have $\hat{p} = .330$ with a sample size $n = 621$; we compute a 95% probability interval for p_B to be (.293, .367). For Gwynn, we have $\hat{p} = .317$ with a sample size $n = 530$; we compute a 95% probability interval for p_G to be (.277, .357).

Let's interpret what these mean. We are 95% confident that Boggs' true batting average (at age 31) is between .293 and .367; that is, the probability that p_B falls between .293 and .367 is 95%. Likewise, we are pretty sure (with confidence .95) that Gwynn's true

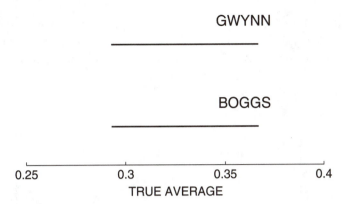

FIGURE 7.2

Display of probability intervals for Gwynn's and Boggs' true batting abilities.

batting average that year p_G is between .277 and .357. We have plotted the two intervals in Figure 7.2.

These two intervals overlap, so it is possible that Gwynn really had a higher batting ability that year and by chance or luck variation, Boggs just happened to perform better. So a 13-point difference in season batting averages is not enough to say that one hitter is better than another hitter. Of course, we know much more about these two hitters than their performance in a single year. In Figure 7.3, we plot the two players' batting averages across all years. (We are plotting AVG against Age.)

This graph is hard to interpret. Why? First, you should notice a lot of up and down variation in both players' batting averages. This is very common. Let me explain why.

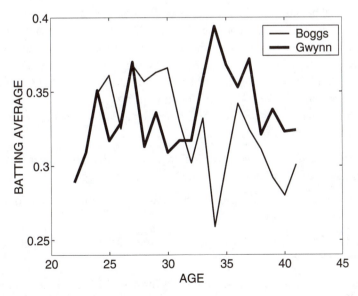

FIGURE 7.3

Graph of Tony Gwynn and Wade Boggs season batting averages against age.

Suppose that we assume that Gwynn is a true $p = .338$ hitter for all of the years of his career (note that .338 is Gwynn's career batting average). I simulated hitting data for Gwynn using his at-bat numbers. Figure 7.4 shows the plot of one simulation.

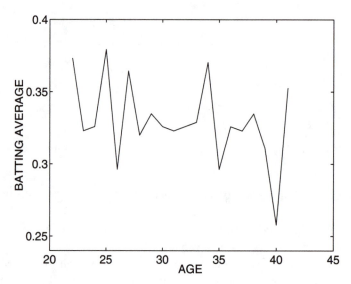

FIGURE 7.4
Simulated batting average for a true $p = .338$ hitter using the at-bat numbers of Tony Gwynn.

This is an interesting plot. We're assuming that Gwynn has the same batting ability across his whole career. His hitting probability is .338 his first year, .338 his second year, etc. But his actual batting performance (his season batting average) shows great variability. The first year he bats .374, the second year he bats .322—this is the natural variation even when Gwynn is assumed to have the same true batting average for all years. So you will typically see a lot of variation in season batting averages for any player. Does this mean that we can't tell if Gwynn or Boggs is the better hitter? No—we can draw a conclusion by looking at the performance of both hitters for all years.

Let's look at Figure 7.3 again. Despite the great up and down fluctuation in the batting averages, we see some general patterns.

- Boggs's season batting averages were high relatively early in his career (pre 30). After 30 he appears to be a weaker hitter despite his good year at age 36.

- Gwynn, in contrast, has maintained a high (around .350) batting average for most of his career.

- Boggs and Gwynn are similar hitters before 32; after 32, Gwynn was consistently higher than Boggs in batting average

Based on the above observations, I think Gwynn was the better hitter on the basis of batting average. You can only make this conclusion based on 20 years of data, not just a single season.

Exercises

Leadoff Exercise. In the following table, some batting statistics for Rickey Henderson for the 1990 and 1991 seasons are displayed.

Season	AB	BB	PA	OBP
1990	489	97		.439
1991	470	98		.400

(a) For each season, compute the number of (approximate) plate appearances by adding at-bats (AB) to walks (BB)—put these values in the table.

(b) Let p_{1990} denote Rickey's true OBP in 1990. Construct a 95% probability interval for p_{1990} using the number of plate appearances (PA) and his on-base fraction OBP.

(c) Construct a 95% probability interval for Rickey's true OBP in 1991, p_{1991}.

(d) Based on your work in (b) and (c), can you say that Rickey's on-base ability was better in 1990 than 1991? Explain how you reached this conclusion.

7.1. Table 7.10 shows the batting average of 12 ballplayers for the years 1998 and 1999.

TABLE 7.10

Batting averages for 12 players for the 1998 and 1999 seasons.

	1998	1999
ABREU	312	335
ALFONZO	278	304
AURILIA	266	281
AUSMUS	269	275
BELLE	328	297
BOONE	266	280
BORDICK	260	277
BROGNA	265	278
BROSIUS	300	247
BURNITZ	263	270
CAIRO	268	295
CANSECO	237	279

(a) Draw a scatterplot of these data, plotting the 1998 AVG on the horizontal axis and the 1999 AVG on the vertical axis.

(b) Comment on any relationship between the 1998 and 1999 AVGs that you see from the scatterplot.

(c) Guess at the value of the correlation coefficient.

(d) The least-squares line to these data is

$$(\text{AVG in 1999}) = 195.0540 + 0.3253(\text{AVG in 1998}).$$

Suppose that a player bats 320 in 1998. Use this line to predict his AVG in 1999.

(e) Explain why there is a positive correlation between a player's 1998 AVG and his 1999 AVG.

7.2. Consider the simple experiment of tossing a fair coin 20 times and recording the number of heads.

(a) What is the probability of a head on a single toss? Why?

(b) Toss a fair coin 20 times and record the number of heads. Put your sequence of Heads and Tails and the number of heads in the following table.

Toss	1	2	3	4	5	6	7	8	9	10	11	12	13	14	15	16	17	18	19	20	Number of Heads
Result (H or T)																					

(c) Combine your result (Number of Heads) with the results of other students in your class. Plot the results using a dotplot on the number line below.

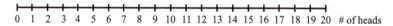

(d) Describe the basic features of the dotplot of number of heads that you constructed in (c).

(e) Suppose that you get seven heads in 20 tosses. Since the fraction of heads is only $\frac{7}{20} = .35$, does that mean that the coin is not fair? Why or why not?

7.3. Suppose that Chipper Jones is really a .300 hitter. That is, his true probability of getting a hit on a single at-bat is .3. We represent his batting ability by means of a spinner shown in Figure 7.5 with a Hit area of .3.

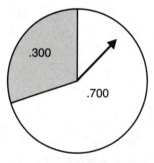

FIGURE 7.5
Spinner to simulate an at-bat for a hitter with a probability of a hit equal to $p = .3$.

Suppose Jones comes to bat 12 times over the weekend. We can simulate 12 at-bats for Jones by spinning the spinner 12 times and counting the number of times the spinner lands in the HIT region. We repeated this process 20 times and below are the number of hits that we observed for these 20 weekends.

4 2 5 8 1 5 5 0 2 4 5 1 2 7 3 2 3 3 2 3

(a) Construct a frequency table of these hit numbers and put your counts in the following table.

Number of hits	0	1	2	3	4	5	6	7	8	9	10	11	12
Count													

(b) What is the most frequent number of hits that Jones gets?

(c) Find the probability that Jones gets five or more hits over the weekend.

7.4. (Exercise 7.3 continued) Suppose that Jones really is a .400 hitter, which means that the chance that he gets a hit is .4. We think of a spinner with a Hit area of .4 (instead of .3 as in the previous exercise) and we simulate the results of 12 at-bats by spinning this .4 spinner 12 times. We repeated this (spinning the spinner 12 times) 20 times, obtaining the following number of hits.

5 3 7 7 1 6 4 3 2 5 2 5 6 3 7 4 6 4 6 4

Construct a frequency table of these hit numbers using the table below and answer questions (b) and (c) from Exercise 7.3.

Number of hits	0	1	2	3	4	5	6	7	8	9	10	11	12
Count													

7.5. (Exercise 7.3 continued) Is it possible to distinguish a true .300 hitter from a true .400 hitter on the basis of 12 at-bats? First, using a spinner, we let a true .300 hitter bat 12 times (a weekend of hitting), and we repeat this simulation for 1000 weekends to get 1000 hit numbers. Likewise, we let a true .400 hitter bat for 1000 weekends, obtaining 1000 simulated hit numbers. We construct count tables in Table 7.11 for the number of hits for both types of hitter.

(a) Find the probability that a true .300 hitter will get exactly five hits during a weekend.

TABLE 7.11
Simulated number of hits in 12 at-bats for true .300 and .400 hitters.

	Number of hits											
	0	1	2	3	4	5	6	7	8	9	10	Total
.300 hitter	17	63	187	224	246	157	69	29	5	3	0	1000
.400 hitter	0	13	69	157	208	216	161	106	45	23	2	1000

(b) Find the probability that a true .400 hitter will get exactly five hits during a weekend.

(c) Suppose that you don't know the batter's ability—he either could be a .300 or .400 hitter. Given that you observe this batter get exactly five hits over the weekend, use Table 7.11 to find the probability that the hitter has a .300 true batting average and a .400 true batting average.

(d) From your computation in (c), are you pretty sure that the hitter has a .400 batting average? Can you really learn much about a batter's ability on the basis of a weekend of hitting (12 at-bats)?

7.6. Suppose that a manager has 11 different types of hitters on his team. One player hits with a true probability of .200, another hits with a true probability of .210, the third player hits with a true probability of .220, . . . , and the last player hits with a true probability of .300. Consider this hypothetical experiment—the manager chooses one player at random from the 11, and then this player comes to bat 20 times. Each time, the manager records

$$p = \text{the true batting average of the player selected,}$$

$$x = \text{the number of hits of this player in 20 at-bats.}$$

Suppose that this hypothetical experiment is repeated 10,000 times, obtaining 10,000 values of p (the true batting average) and x (the number of hits). These values are organized by means of the two-way table shown in Table 7.12.

TABLE 7.12

Two-way table of simulated values of true batting average and number of hits in simulation.

		\multicolumn{12}{c}{TRUE BATTING AVERAGE (p)}
		.200	.210	.220	.230	.240	.250	.260	.270	.280	.290	.300	All
	0	8	6	6	5	3	5	1	2	3	3	4	46
	1	71	38	39	27	24	20	17	14	14	8	5	277
	2	107	109	85	84	65	51	54	45	40	37	25	702
#	3	209	169	150	164	119	137	120	94	99	77	75	1413
	4	192	213	167	205	179	148	165	132	135	146	108	1790
O	5	173	176	191	186	181	175	181	184	166	152	140	1905
F	6	87	121	143	143	137	147	154	163	167	185	165	1612
	7	45	64	79	79	83	97	102	129	156	134	126	1094
H	8	14	23	38	32	52	54	84	78	84	113	100	672
I	9	7	4	7	17	21	29	34	30	38	52	57	296
T	10	3	3	4	3	8	13	9	15	9	37	23	127
S	11	0	0	2	1	5	3	4	5	6	10	11	47
(x)	12	0	0	1	1	0	0	1	2	2	1	6	14
	13	0	0	0	0	0	0	0	1	0	2	2	5
	All	916	926	912	947	877	879	926	894	919	957	847	10000

(a) How many times was a player with a .22 ability chosen and four hits were observed? Find the probability that a player with a .22 ability is chosen and four hits are observed.

(b) What is the most likely number of hits observed? What is the probability of this number of hits?

(c) Suppose that a hitter only gets two hits (out of 20). What is the chance that he is a .200 ability hitter?

(d) If a hitter only gets two hits, find the probability that the hitter's ability (p) is at most .25.

(e) If a hitter gets two hits, find the smallest interval of values that contains the ability p with a probability at least .9.

7.7. (Exercise 7.6 continued)

(a) Suppose that the player gets five hits. Use the two-way table to find the probability that the player has a true batting average (p) greater than .25, and the probability that the player has a true batting average .25 or smaller. Put your answers in the table below.

Observe 5 hits in 20 at-bats

Event	Probability
Player's true average is greater than .25	
Player's true average is .25 or smaller	

(b) Repeat (a) assuming that the player gets 8 hits.

Observe 8 hits in 20 at-bats

Event	Probability
Player's true average is greater than .25	
Player's true average is .25 or smaller	

(c) Compare your answers to parts (a) and (b). In which case (observing five hits or observing eight hits) did you learn more about the player's true batting ability? Why?

7.8. Suppose Moises Alou has 50 at-bats at some point in the 2003 season and has 14 hits.

(a) Find Alou's current batting average (his average over the 50 at-bats).

(b) Find a 95% probability interval for Alou's true batting average for the 2003 season.

(c) Is it possible that Alou has a true batting average of .320? Is it likely? Why?

7.9. Suppose you are following Todd Helton's batting average during the 2003 season.

(a) Suppose that Helton has 40 hits after 100 at-bats. Find a 95% probability interval for Helton's true 2003 batting average.

(b) Suppose that Helton has 80 hits after 200 at-bats. Find a 95% probability interval for Helton's batting average.

(c) Compare the two probability intervals you computed in (a) and (b) with respect to the center and length of the interval.

7.10. In 1999, Larry Walker had 166 hits in 438 at-bats, and Tony Gwynn had 139 hits in 411 at-bats.

(a) Find the 1999 AVGs for Walker and Gwynn.

(b) Find 95% probability intervals for Walker and Gwynn's true batting averages.

(c) Are you confident that Walker really was a better hitter than Gwynn during the 1999 season? (Use the probability intervals you calculated in (b) to answer the question.)

Further Reading

Albert and Bennett (2001), Chapter 3, introduce statistical inference in the context of baseball. A spinner is used to model a player's hitting ability, and they use a simulation, such as described in Case Study 7-3, to learn about a player's ability based on his batting performance. Introductory inference is described from a Bayesian perspective in Berry (1996) and Albert and Rossman (2001). Basic inferential methods for one proportion are contained in Devore and Peck (2000) and Moore and McCabe (1998).

8

Topics in Statistical Inference

What's On-Deck?

In this chapter, we focus on two interesting statistical inferential topics related to baseball—the interpretation of situational data and the search for true streakiness in baseball data. Today baseball hitting and pitching data is recorded in very fine detail, and it is popular to report the performance of hitters and pitchers in a large number of situations. For example, we record how a player hits in home and away games, against left- and right-handed pitchers, during different months of the season, during different pitch counts, and against different teams. The reporting of this situational data raises an interesting question: how much of the variation in this data corresponds to real effects and how much of the variation is attributed to luck or chance variation? In Case Study 8-1, we look at the situational hitting data that is reported for a single player. When we look at the situational hitting data for the home vs. away situation for a number of players (Case Study 8-2), we will see some interesting effects. Some players will hit for a much higher average during home games and other players perform much better during away games. But when we graph the situational effects for a group of players during two consecutive years, we will see that there is no association. In other words, players don't appear to possess an ability to perform unusually well (or poorly) during home games.

In Case Studies 8-3 and 8-4, we describe some useful statistical models for situational data. We can represent the hitting abilities of players by means of a normal or bell-shaped curve. Many situations are in the "no effect" scenario—here the player has the same probability of getting a hit in either situation. Other situations are so-called biases—the situation, such as playing at home, will add a constant number to every player's hitting probability. The most interesting situation can be regarded as an "ability effect" where the particular situation is to one player's advantage, and to another player's disadvantage. Generally speaking, most of the variation in the reported situational data is essentially noise or chance variation, and it is difficult to pick up real situational effects in baseball data.

In the last two case studies, we look at the general topic of streakiness in baseball hitting data. In Case Study 8-5, we look at a player, John Olerud, who is generally believed to be a streaky hitter, and discuss ways of measuring streakiness in his day-to-day hitting data. By looking at some of these streaky statistics, there may be some evidence that Olerud is genuinely a streaky hitter. But Case Study 8-6 shows that these patterns of streakiness are also common in results in tossing a coin many times. The conclusion from this brief study is that genuine streakiness in hitting data is difficult to detect.

Case Study 8-1: Situational Hitting Statistics for Todd Helton

Topics Covered: Introduction to situational hitting data   In this chapter, we first focus on situational statistics. These are currently very popular among baseball fans. The book *STATS Player Profiles 2001* contains one of the best collection of hitting and pitching situational stats and it's fun reading.

When you watch a baseball game, you'll hear the announcer say something like "Scott Rolen has a .421 batting average when he is facing Greg Maddux at Veteran's Stadium." You are supposed to be surprised by this statement. If you are watching Rolen bat against Maddux in Veteran's Stadium, you might expect Rolen to get a hit.

But is this the right interpretation? How can we make sense of all of these interesting situational stats that we hear in the media?

Let's focus on Todd Helton, the Rockies hitter, who had a great 2000 season. Here are Helton's basic hitting stats.

	AVG	AB	OBP	SLG
Total	.372	580	.463	.698

Next, we see how Helton did against left-handed pitchers and right handers. Generally a hitter bats better against a pitcher who throws with an arm opposite from which he takes his batting stance. Managers believe in this effect and make substitutions based on this belief. As expected, Helton, a left-handed hitter, hit better against right-handers.

	AVG	AB	OBP	SLG
vs. Left	.329	143	.451	.594
vs. Right	.387	437	.467	.732

Next, we see how Helton batted at home and on the road. Generally it is believed that ballplayers perform better at home (more comfortable surroundings, loving fans, home cooking, etc.) Helton did bat better at home—part of this is likely the Coors Field effect. The Rockies' ballpark is well-known as friendly to hitters.

	AVG	AB	OBP	SLG
Home	.391	302	.484	.758
Away	.353	278	.441	.633

The next situation refers to the runners on base. "None on" means the bases are empty. We see that Helton was a better hitter (actually performed better) with runners on base.

	AVG	AB	OBP	SLG
None on	.340	294	.409	.684
Runners on	.406	286	.512	.713

Next we see how Helton hit each month of the baseball season. There is quite some variation here. He was hot in May and August (I call the averages 512 and 476 WOW statistics) and cold in September. How do we interpret these numbers? Was Helton a much better hitter in August than in September?

	AVG	AB	OBP	SLG
April	.337	89	.440	.663
May	.512	82	.588	1.000
June	.315	92	.422	.565
July	.340	106	.427	.547
August	.476	105	.548	.848
September	.267	101	.369	.594
October	.400	5	.400	1.000

The next situations are runner and out situations. None on/out means there are no outs and no runners on base. Scoring position means that there is at least one runner on 2nd or 3rd. ScPos/2 Out means runners in scoring position and 2 outs—this is a pressure situation. Helton was relatively weak (321) here, but it is based on only 56 at-bats.

	AVG	AB	OBP	SLG
None on/out	.314	140	.377	.621
Scoring Posn	.392	153	.525	.791
ScPos/2 Out	.321	56	.558	.714

Close & Late means that you are in a late inning (say 7th or later) and the batting team either is ahead by one run, tied, or with the potential tying run at least on deck. This is also a clutch situation. Helton was especially good (583) with the bases loaded, but he only had 12 at-bats in this situation.

	AVG	AB	OBP	SLG
Close & Late	.393	84	.495	.774
Bases Loaded	.583	12	.647	.750

Batting #3, etc. refers to the position in the batting order. Helton batted mainly in the 4th and 5th positions, and was strong both places in the order.

	AVG	AB	OBP	SLG
Batting #3	.311	90	.385	.511
Batting #4	.391	243	.497	.770
Batting #5	.380	245	.459	.702
Batting #6	.000	1	.000	.000
Batting #9	.000	1	.000	.000

Next, we see how Helton batted playing different fielding positions. This is not at all informative, since he played practically all of his games at first base.

	AVG	AB	OBP	SLG
As 1b	.374	578	.465	.701
As PH	.000	2	.000	.000

Next, we see how Helton hit during different pitch counts. Generally players bat better when they have a pitch advantage (like 1-0, 2-0 or 3-0), and bat much worse when there are two strikes (the pitcher has the advantage). Helton's averages are consistent with this pattern.

	AVG	AB	OBP	SLG
0-0 count	.375	96	.479	.656
After (0-1)	.357	238	.399	.630
After (1-0)	.386	246	.511	.780
Two strikes	.336	268	.393	.571

Next, we see how Helton batted on fields with natural grass and on artificial turf fields. The batted ball behaves differently on the two surfaces and this can affect batting averages.

	AVG	AB	OBP	SLG
Grass	.367	521	.457	.687
Turf	.424	59	.520	.797

How did Helton bat during day games and night games?

	AVG	AB	OBP	SLG
Day	.391	215	.470	.744
Night	.362	365	.459	.671

How did he bat during the early innings (1–6) versus the late innings?

	AVG	AB	OBP	SLG
Inning 1-6	.379	398	.462	.696
Inning 7+	.357	182	.466	.703

How did he bat against different teams?

	AVG	AB	OBP	SLG
vs. Ana	.300	10	.364	.400
vs. Mil	.351	37	.419	.649
vs. Oak	.538	13	.625	1.154
vs. Sea	.182	11	.308	.182
vs. Tex	.231	13	.286	.308
vs. Atl	.294	34	.415	.559
vs. ChC	.114	35	.244	.229
vs. Cin	.257	35	.278	.629
vs. Hou	.481	27	.625	1.185
vs. LA	.255	47	.351	.383
vs. Mon	.625	32	.683	1.125
vs. NYM	.478	23	.586	1.174
vs. Phi	.531	32	.583	.844
vs. Pit	.500	32	.550	.875
vs. StL	.400	25	.543	.880
vs. SD	.396	48	.464	.646
vs. SF	.404	47	.525	.638
vs. Fla	.457	35	.500	.914
vs. Ari	.273	44	.347	.545

How did he bat before the All-Star game and after the All-Star game?

	AVG	AB	OBP	SLG
Pre-All Star	.383	298	.479	.701
Post-All Star	.362	282	.446	.695

To sum up, although Helton overall was a .372 hitter in the 2000 season, he appears to bat much better (or much worse) in particular situations. Specifically, we see that Helton

- batted 58 points better against left-handed pitchers,

- batted 38 points better at home games,

- batted for a high average in May and August and for a low average in September,

- batted for a .583 average when bases were loaded.

The main question we will address in the following case studies is: Do these observed situational effects correspond to real effects?

For example, can we say that there is really an advantage to hitting in Coors Field? Maybe Helton has the same ability to get a base hit in home and away games and, by luck or chance variation, he happened to hit better at home this year. Or maybe it really is easier to get base hits in Coors Field due to "better" weather conditions, and Helton's difference in batting averages is just a reflection of this real Coors Field effect. We will see in this chapter that much of the variation that we see in situational data such as Helton's batting averages can be explained by chance, and it is relatively difficult to pick out real situational effects.

Case Study 8-2: Observed Situational Effects for Many Players

Topics Covered: Stemplot, scatterplot, relationship between two variables, distinction between ability and performance   To get an understanding of the patterns of situational data, we look at Table 8.1 where we see the home batting average and away batting average for a group of 20 hitters for the years 1999 and 2000. Let's focus first on the 2000 data. For each player, we compute the difference

$$\text{DIFF} = \text{Home AVG in 2000} - \text{Away AVG in 2000}$$

TABLE 8.1

Home and away batting averages for two years for 20 players.

Name	2000			1999		
	Home	Away	Diff	Home	Away	Diff
Alex Rodriguez	.272	.356	−.084	.284	.286	−.002
Albert Belle	.273	.288	−.015	.304	.289	.015
Barry Bonds	.321	.291	.030	.247	.276	−.029
Chipper Jones	.323	.299	.024	.366	.275	.091
Tito Martinez	.304	.341	−.037	.360	.308	.052
Todd Helton	.391	.353	.038	.385	.252	.133
Derek Jeter	.338	.340	−.002	.329	.369	−.040
David Justice	.291	.281	.010	.305	.267	.038
Ken Griffey	.289	.254	.035	.288	.283	.005
Ken Caminiti	.242	.350	−.108	.295	.276	.019
Jeff Kent	.335	.333	.002	.246	.332	−.086
Kenny Loften	.274	.286	−.012	.272	.330	−.058
Manny Ramirez	.357	.345	.012	.311	.354	−.043
Nomar Garciaparra	.375	.370	.005	.378	.331	.047
Rafael Palmeiro	.292	.285	.007	.325	.323	.002
Roberto Alomar	.314	.306	.008	.333	.313	.020
Scott Rolen	.327	.274	.053	.266	.271	−.005
Sammy Sosa	.306	.332	−.026	.325	.252	.073
Frank Thomas	.347	.309	.038	.310	.299	.011
Jim Thome	.309	.230	.079	.282	.273	.009

For example, Alex Rodriguez batted .272 at home games and .356 at away games for a difference of DIFF = .272 − .356 = −.084. (Rodriguez actually batted much better in away games in the 2000 season.) We plot the differences for all 20 players using a dotplot in Figure 8.1.

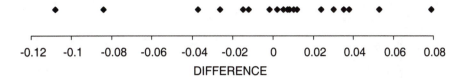

FIGURE 8.1

Dotplot of differences in home and away batting averages for 20 players in the 2000 season.

What can we see in this dotplot?

- We can compute the median difference; it is .0075. This means, on average, that a hitter bats 7.5 points higher during home games than during away games. This is what we expected above.

- However, there is a tremendous range in these difference values. One hitter had a −.108 difference—this player actually batted 108 points better in away games. In contrast, another hitter had a difference of .079. This player batted 79 points higher at home. Generally, situational data is interesting since one will typically see a great range of differences—there will be many large positive values and large negative values.

Since we see many "extreme" home/away effects, it is tempting to try to explain why a particular player is doing so well (or so poorly) at home. But do players really have different *abilities* to use the home field advantage? Is it possible that one player uses the home field to great advantage in his hitting, while a second player has the same batting ability at home and away games? We can learn about the presence of home/away batting abilities by looking at data for many players.

Demonstration That Players Have Different Batting Abilities

It is obvious to most baseball fans that players have different batting abilities. But how can we demonstrate this fact from hitting data? Let's look at the home hitting data in Table 8.1. For each player, we collect his 1999 home batting average and his 2000 home batting average. Figure 8.2 displays a scatterplot of the 1999 and 2000 home averages for the 20 players. Looking at the graph, we see that the points drift from the lower left to the upper right regions—this means that the 1999 and 2000 batting averages are positively associated. (The correlation between the two variables is .44.) This means that players who hit well at home in 1999 tend also to hit well at home the following year. Likewise, a weak hitting season by a player in 1999 tends to be associated with a weak-hitting performance the following season. The explanation for the positive association in the graph is that players have different batting abilities—the better hitters correspond to points in the upper right section of the graph and the weak hitters correspond to points in the lower left section.

Do Players Have Different Home/Away Abilities?

We use the above strategy to search for situational abilities. Let's consider Pat Burrell, who had a good rookie season with the Phillies. Suppose that this year Pat bats 50 points better

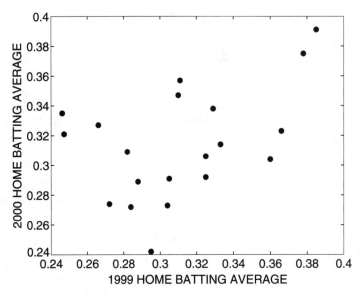

FIGURE 8.2
Scatterplot of 1999 and 2000 home batting averages for 20 players.

at home than away games. Does Pat really have an extra ability to hit well during home games? We can see if this is a real effect by looking at next year's performance. If Burrell continues to hit much better at home, then we would be more confident that he has some extra home batting ability. If we have home/away batting data for a group of players for two consecutive years, we look for a general relationship between a player's home/away effect one year with the corresponding effect the next year.

In Table 8.1, we have computed the difference between the home and away batting averages (DIFF) for both the 1999 and 2000 baseball seasons for our 20 players. To see if there is a general relationship between a player's 1999 DIFF and his 2000 DIFF, we construct a scatterplot in Figure 8.3.

In contrast to the pattern in Figure 8.2, we don't see any trend in this graph. (The correlation between the two variables in this case is .04.) There does not appear to exist a tendency to show the same type of home/away effect for two consecutive years. This suggests that the home/away effect is not an ability characteristic. This means that a player who hits unusually well at home one particular year generally will not hit unusually well at home the next year—likewise poor home hitters one year will not be poor hitters at home the following year. Remember that there will be a positive association between a player's batting average (or any other batting measure) for two consecutive years—this means that players have intrinsic batting abilities. But there is no general tendency for players to hit better (or worse) in the home/away situation for two consecutive years.

Case Study 8-3: Modeling Batting Averages for Many Players

Topics Covered: Probabilities of hitting for many players, a random effects model, assessing the goodness of fit of a model, normal distribution   In our earlier case studies:

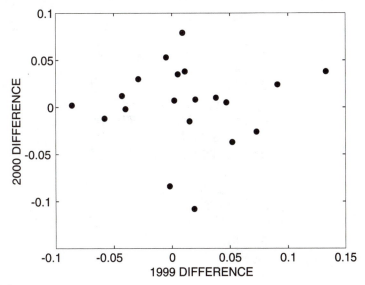

FIGURE 8.3
Scatterplot of differences in home and away batting averages for 20 players in 1999 and 2000 seasons.

- We distinguished between the *ability* of a hitter and his *performance*.

- We observed a hitter's performance during the season; we don't know a batter's ability, but we'll learn about it from the hitting data.

- We modeled a hitter's ability by a spinner. The spinner has two areas, HIT and OUT, and the area of the HIT region is given by p. This is our measure of ability.

Now instead of one player, we consider all 119 players who were "regulars" (had at least 300 at-bats) in the American League in 2001. Figure 8.4 shows a stemplot of their batting averages.

We see that the batting averages are bell-shaped about the mean value of .274. The highest batting average was Ichiro Suzuki at .350 and the lowest was Brady Anderson at .202. Can we construct a model for the true batting averages (the abilities) for these 119 hitters?

The simplest model I can think of is what we call the "One Spinner" model. Maybe all of the 119 players have the same batting ability. Each player is using the same spinner with hitting probability $p = .274$ (the average of all players). If you believe this, then the variation in the season batting averages that we observe in the stemplot above are simply due to chance variation. Players all have the same ability, but some, like Suzuki, were lucky and had high season averages; others, like Anderson, were unlucky and had low averages.

Now, if you know anything about baseball, you should be thinking that this is a crazy model—players do have different batting abilities. In fact, we illustrated this fact in the last case study where we looked at batting averages of some players for two consecutive seasons. I agree, but I want to demonstrate how a statistician checks the suitability of a probability model.

20	2
21	
22	067
23	233489
24	0122589
25	000001233456677889
26	0112333345567777778899
27	0012334446677789
28	0000002233456677899
29	11338
30	122345666678
31	11688
32	5
33	016
34	2
35	0

FIGURE 8.4
Stemplot of the batting averages of all American League players with at least 300 at-bats in the 2001 season.

Simulating Hitting Data Using a One Spinner Model

To see if the "One Spinner" model is reasonable, we simulate hitting data from the model. We imagine 119 spinners, each with hitting probability $p = .274$. We spin each spinner for a whole season using the same numbers of at-bats as the 2001 players and obtain 119 season batting averages. We do this one time. In the back-to-back stemplot display in Figure 8.5, we have placed the simulated batting averages on the left and the actual 2001 batting averages on the right.

How does the simulated data compare with the actual data? There is a substantial difference—the simulated batting averages appear to have less variation or spread than the real averages.

We can measure spread of a dataset by the standard deviation s. For the dataset of real batting averages, we compute $s = .0268$ and for the simulated data, $s = .0199$. This confirms what we just said—the simulated averages aren't as spread out as the real averages.

To see if this always happens, we did the simulation from the "One Spinner" model many times. In four additional simulation runs, we obtained

$$s = .0199, \quad s = .0180, \quad s = .0206, \quad s = .0207.$$

Each time, the standard deviation is smaller than the standard deviation (.0268) for the real data. So this model does not seem to generate data that is similar to the 2001 dataset of averages. We conclude the "One Spinner" model isn't appropriate and consequently that batters have different abilities.

	AVGS FROM ONE-SPINNER MODEL		ACTUAL AVGS
		20	2
		21	
	7	22	067
	8883311	23	233489
	999887761	24	0122589
	98877754443321	25	000001233456677889
	88644332221	26	01123333455677777778899
	98887776666665544433333332221100	27	0012334446677789
	9998776665554443332110	28	0000002233456677899
	88887555432211	29	11338
	652211111	30	122345666678
	1	31	11688
		32	5
		33	016
		34	2
		35	0

FIGURE 8.5
Back-to-back stemplots of simulated batting averages from one-spinner model and 2001 American League batting averages.

A Many Spinners Model

What is an alternative probability model that can represent hitting data in baseball? A "Many Spinners" model seems to better represent what is really going on. We first represent the hitting abilities of the 119 players by a normal curve shown in Figure 8.6 with an

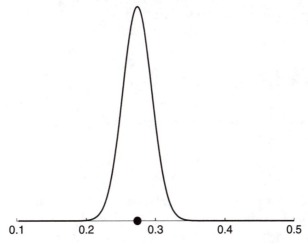

FIGURE 8.6
Bell-shaped curve to represent the hitting probabilities of the American League players.

average of .274 and standard deviation of .021. In other words, the hitting probabilities of the players are variable according to a normal curve with mean .274.

We sample at random 119 hitting probabilities from this normal curve. Then we represent the abilities of the players using a set of spinners, where the HIT areas are different for different players. One player might have a hitting area of $p = .282$, another may have a hitting area of .304, and so on.

We tried simulating data using this "Many Spinners" model. We first simulated a set of random hitting probabilities and then simulated hitting data using these probabilities. Figure 8.7 shows back-to-back stemplots of the simulated season batting averages and the actual 2001 batting averages.

AVGS FROM MANY-SPINNERS MODEL		ACTUAL AVGS
3	18	
	19	
	20	2
004	21	
986	22	067
86533	23	233489
54322100	24	0122589
98766664321	25	000001233456677889
9877666664443333200	26	011233334556777778899
99988866654444431110	27	0012334446677789
98877665555444322110	28	0000002233456677899
97776531100	29	11338
99765210	30	122345666678
7650	31	11688
8330	32	5
1	33	016
3	34	2
	35	0

FIGURE 8.7

Back-to-back stemplots of simulated batting averages from many-spinners model and 2001 American League batting averages.

Comparing the two stemplots, the distribution of the simulated data from the "Many Spinners" data does resemble the distribution of the actual 2001 batting averages. In particular, the standard deviations match up—the standard deviation of the simulated data is .0272, which is close to the standard deviation of the observed data .0268. So the "Many Spinners" model seems to be a good representation for the hitting abilities of many players.

We will use this probability model in the following case study when we model situational hitting data.

Case Study 8-4: Models for Situational Effects

Topics Covered: Probability models for situational hitting data, bias model, model with ability effects   What are good models for situational data? We describe three basic models here that seem to describe the pattern in hitting data for all of the situations that are discussed in the *STATS Player Profiles* book.

Recall our basic probability model for batting data for a group of players described in Case Study 8-3. Players have different abilities (values of p) selected from a normal curve (with a mean of .274 and a standard deviation of .021), and batting data is found by spinning a bunch of spinners, where each spinner has a different hitting probability.

The No Effect Model

To describe the simplest situational hitting model, suppose we have five spinners with hitting probabilities of $p = .2, .3, .4, .5, .6$. These spinners represent the abilities for five baseball players. Suppose that we spin the spinners in the dark and the light. Now, it is reasonable to think that the lightness of the room has no effect on the probability of getting a hit on a particular spinner. So if the first spinner has a hitting probability of $p = .2$ in the light, then this spinner will also have a hitting probability of $p = .2$ in the dark. Likewise, the .3 spinner will have a hitting probability of .3 both in the light and the dark, the .4 spinner will have the same hitting probability in the dark and in the light. In this case there is no true situational effect—so we call this the "no effect" model.

Now suppose we spin the .2 spinner 100 times in the dark and 100 times in the light. Now it is certainly possible that we'll get 22 hits in the dark and only 18 hits in the light for an observed situational effect of

$$\text{Observed Situational Effect} = 22/100 - 18/100 = .04$$

But this effect is just due to chance—there is no true situational effect. An example of this "no effect" situation is before and after the All-Star Game. Generally, players appear to have the same batting abilities before and after the All-Star Game. Certainly, we will observe some players one season who bat better before the All-Star Game and we'll see other players who are hot after the All-Star Game. But most of the variation in the differences in batting averages before and after the All-Star Game is due to chance variation.

The "Bias" Model

The bias model is a different way of representing the hitting abilities of players in two situations. Here the situation changes the hitting probability by the same amount for all hitters.

Let's consider our dark and light example again. Let's suppose that it is hard to see the spinner in the dark, and the light increases the hitting probability by .05 for all hitters. So if one hitter has a dark hitting probability of $p_{dark} = .2$, it will be $p_{light} = .2 + .05 = .25$ in the light. Another hitter with a dark hitting probability of $p_{dark} = .4$ will have a hitting probability of $p_{light} = .45$ in the light.

The bias model seems appropriate for the home vs. away hitting data. A particular ball-park has a positive or negative impact on a player's batting average. Coors Field is an obvious example of a ballpark that helps a player's batting ability, and Dodger Stadium is an example of a ballpark that hurts a player's batting ability. But the effect of the ballpark is to add a constant number to each player's batting probability. So Coors Field may add a positive number, say .10 to each player's true batting average. There is a situational effect here, but this effect is the same for all players.

A Simulation

I used a statistics computing package to illustrate what situational data looks like if there is a bias effect. I assumed

- there are 100 players

- each player has 300 at-bats at home and 300 at-bats away,

- there is a bias situational effect of ten points due to playing at home—so each player's home hitting probability is .010 higher than his away hitting probability.

The program first simulated 100 away true batting averages from a normal curve with mean .274 and standard deviation .021—these hitting probabilities are put in the p_AWAY column. We compute home true batting averages in the p_HOME column by adding .010 to each probability in the p_AWAY column. We then simulated hits for 300 at-bats at the home games using the home hitting probabilities—these numbers are put in the h_HOME column. Similarly, we simulated hits for 300 away at-bats using the away hitting probability—these hit numbers are in the h_AWAY column. We computed batting averages at home and away (these AVG's are in the AVG_HOME and AVG_AWAY columns), and computed observed differences—batting average (home) - batting average (away). A part of the output of this simulation is shown in Figure 8.8.

	p_AWAY	p_HOME	h_AWAY	AVG_AWAY	h_HOME	AVG_HOME	DIFF
1	0.27551	0.28551	78	260	96	320	60
2	0.252422	0.262422	80	267	75	250	−17
3	0.276173	0.286173	90	300	89	297	−3
4	0.259321	0.269321	81	270	92	307	37
5	0.301628	0.311628	91	303	83	277	−26
6	0.292472	0.302472	94	313	90	300	−13
7	0.262338	0.272338	78	260	90	300	40

FIGURE 8.8
Some situational hitting data assuming that the home ballpark adds .010 to the probability of getting a hit for all players.

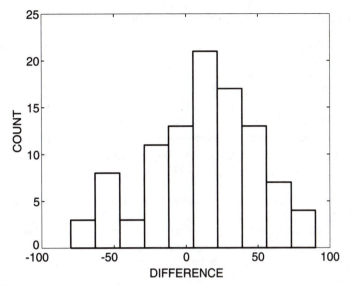

FIGURE 8.9
Histogram of observed situational effects from simulated data where there is a bias situational effect.

Figure 8.9 displays a histogram of the observed situational effects contained in the DIFF column. Here we see that when there is a bias situational effect, the observed effects can look interesting. One player in the simulation hit for 80 points better at home games and another hit for 73 points higher away. This high variation is misunderstood by baseball fans. People think that some players (like the one that hit 80 points higher during home games) have a special ability to play better at home, when really the home ballpark has the same positive effect on all players.

The "Ability" Model

The last model, the so-called "ability model," is the most complicated description of situational data. This probability model says that there are real situational effects, and they are different among players. I'll illustrate this scenario using the following "Joe Cool" and "Harry Hyper" example—two players who react differently to a home crowd.

Joe Cool plays the same way everywhere he plays. So if Joe hits for a batting probability of .300 at home, he'll also hit for a batting probability of .300 on the road. Harry, in contrast, is very emotional (like Pete Rose) and feeds off of the enthusiasm of the home crowd. If Harry hits for a probability of .275 on away games, he might hit 50 points higher, or .325, at home.

There are relatively few situations (among the ones that are discussed) that are ability effects. The one exception is the batting performance of the hitter under different pitch counts. Suppose we look at two batting averages:

- the batting average when there are two strikes on the batter,

- the batting average when the batter is ahead in the count (that is, the pitch count is 3–0, 3–1, 2–0, 2–1, or 1–0).

If we look at the situational hitting data for all players in this situation, there is evidence that pitch count is an ability effect. This means that players have different abilities under various pitch counts. The big slugger (Jim Thome comes to mind) is an ineffective hitter when he is behind in the count. When the count is 0 balls and 2 strikes, he's likely to strike out. In contrast, the good contact hitters (like Tony Gwynn) are effective hitters even when they are behind in the count. Suppose we define the pitch count situational effect to be

$$\text{DIFF} = \text{AVG(ahead in the count)} - \text{AVG(behind in the count)}.$$

There is evidence to suggest that big sluggers (like Thome) have large values of this situational effect, and other hitters (the contact-hitter like Gwynn) have relatively small pitch count effects.

How do we find a group ability effect in situational data? Recall our discussion about looking for abilities in hitting data in Case Study 8-2. Suppose you look at a group of players and compute their situational effect (say, batting average ahead in the count minus batting average behind in the count) for two consecutive years, say 1999 and 2000. Construct a scatterplot of the two years of situational effects. You have found a group ability situational effect if you find a positive association in the scatterplot between the 1999 effect and the 2000 effect. This means that players with high situational effects in 1999 will tend to have high situational effects in 2000. Also players who have low situational effects one year will tend to have low effects the second year. Remember that most situational effects in baseball hitting are "no effects" or "biases"—the pitch count is one of the few situations where players generally appear to have different abilities to take advantage of the situation.

Case Study 8-5: Is John Olerud Streaky?

Topics Covered: Distinction between streaky performance and streaky ability, moving averages, runs   In this case study we talk about streakiness or the hot hand.

Do players really get hot and cold? Here I'm not talking about the hot and cold streaks that we observe in baseball data. I'm instead talking about a player's ability to "get into a groove" in hitting or pitching. Maybe during a certain week, the player's batting stroke feels just right or he sees the ball particularly well. Another week, he feels different— maybe his batting stroke is out of sync or he has an injury that makes hitting more difficult. Is there evidence for a player to be truly streaky, that is, have an ability to be hot and cold?

I wanted to find a baseball player who is generally perceived to be streaky. I went to the *Google* search engine and did a search for "streaky hitter." I found an article about John Olerud in the `www.metclubhouse.com` website. A paragraph from the article is shown below.

This quality causes many people to stamp Olerud with the label of "consistency" (meaningless subjective buzzword #571). In fact I am right now looking at a Mets

Magazine graced by John's photo on the cover with the headline "Mr. Consistency." It strikes me as a strange way of describing a player who, in 9 seasons, has hit .363 once, .354 once—and .297 or lower the other seven times. In fact, watching him now for three years, I would characterize Olerud as a fairly streaky hitter prone to torrid hot streaks and horrible slumps several times a year. In this regard he reminds me of another Met, who incidentally also came to the club in mid-career, from a small market, amid much inane speculation about whether he could "handle the pressure of playing in New York": Kevin McReynolds. Both McReynolds and Olerud have been stamped by the press as the "consistent" types, more because of their personalities and their images than because of anything to do with their actual play (McReynolds was also a very streaky hitter).

Here's an introduction to John Olerud. He currently plays first base for the Mariners. He started with the Blue Jays, was traded to the Mets, and was recently signed by the Mariners. It is interesting to look at Olerud's batting averages for the 12 seasons 1990–2001—a stemplot of these averages is shown in Figure 8.10.

```
25 │ 6
26 │ 5
27 │ 4
28 │ 45
29 │ 71489
30 │ 2
31 │
32 │
33 │
34 │
35 │ 4
36 │ 3
```

FIGURE 8.10
Stemplot of batting averages for John Olerud for the seasons 1990-2001.

There is an interesting pattern here. Most of his season averages are in the .280–.300 range, and he had two great seasons where he hit in the .350–.360 range. I think his current true batting probability is likely in the .280–.290 range, although I would be interested in explanations for his two great years. To look for streakiness in Olerud's hitting data, we focus on his game-to-game hitting data for the 1999 season (the last year he played for the Mets).

Moving Average

One way of looking for streaky behavior is to compute a *moving average*. This is a short-term batting average using a window of a particular number of games. Suppose that we wish to compute moving averages using a window of five games.

1. The first moving average is the batting average for games 1 through 5.

2. The second moving average is the batting average of games 2 through 6.

3. The third moving average is the batting average of games 3 through 7.

Essentially we're just computing averages over short time intervals. In Table 8.2, I compute moving averages for the 1999 Olerud data using a window of 9 games.

TABLE 8.2
Computation of moving averages for day-by-day batting data for John Olerud.

Game	Date	Opp	AB	H			
1	4/5/99	@Fla	4	2			
2	4/6/99	@Fla	4	1			
3	4/7/99	@Fla	4	1			
4	4/8/99	@Mon	2	0			
5	4/9/99	@Mon	3	0	33	11	0.333
6	4/10/99	@Mon	5	3	31	9	0.290
7	4/11/99	@Mon	5	4	31	10	0.323
8	4/12/99	Fla	3	0	30	10	0.333
9	4/14/99	Fla	3	0	30	12	0.400
10	4/15/99	Fla	2	0	31	13	0.419
11	4/16/99	Mon	4	2	27	10	0.370
12	4/17/99	Mon	3	1	26	8	0.308
13	4/18/99	Mon	2	2	27	9	0.333
14	4/20/99	@Cin	4	1	28	10	0.357
15	4/21/99	@Cin	1	0	28	10	0.357

1. In games 1–9, Olerud had 33 AB and 11 H for an average of $\frac{11}{33} = .333$. This moving average of .333 is put in the table across the average game number 5 (5 is the average of games $1, 2, \ldots, 9$).

2. In the table, we see a moving average of .419 across game 10. We look at the nine games centered about game 10—these are games 6 through 14. In this period, Olerud had 13 hits in 31 at-bats for a batting average of $\frac{13}{31} = .419$.

3. To get a moving average of .413 for game 17, we look at the nine games about game 17—these are games 13 through 21. He was 12 for 29 in this period for a moving average of $\frac{12}{29} = .413$.

If we graph the game numbers against the corresponding moving averages, we get the graph shown in Figure 8.11. This graph is a picture of the hot and cold periods of Olerud's hitting for the 1999 season. We see

• a hot period about game 45—during one nine game period, Olerud batted .500,

• this hot period was surrounded by two cold periods, where he batted close to .200,

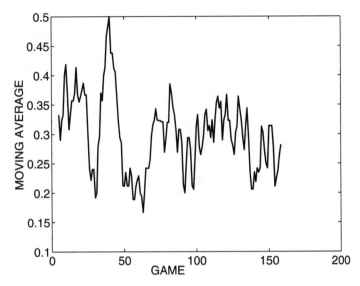

FIGURE 8.11

Moving average plot of John Olerud's batting average using a window of nine games.

- Olerud's hitting was more consistent the last half of the season—we see fewer hills and valleys.

So one indication of streakiness is the extreme hills and valleys that we see in the moving average plot.

Runs

Another way of measuring streakiness is based on runs of good and bad days (not to be confused with runs scored in a baseball game).

Suppose we classify each day of Olerud's hitting as being either H (hot) or C (cold)—we say he is hot if his day's batting average is over .300 (his season avg); otherwise he is classified as cold. Table 8.3 shows Olerud's batting results for the first 13 games and the hot and cold classification.

In Game 1, Olerud was 2/4 = .500; this is larger than .300, so we call it a Hot (H) day. In Game 2, he was 1/4 = .250 which is under .300, so he is Cold (C) that day.

Now we look for runs in this sequence. A **run** here is simply a consecutive sequence of Hs or a sequence of Cs. Specifically, we count the total number of runs, and the length of the longest run (either H or C).

In the above sequence of 13 games, we see that total number of runs = 5, and longest run is CCCC, so length of longest run = 4.

If the player is streaky, we expect to see a small number of runs, and long runs of Hs or Cs, so the length of the longest run would be *large*. This makes sense. If a player is streaky,

TABLE 8.3

Batting results for John Olerud for the first 13 games of the 1999 season and classification of the result in hot and cold states.

Game	Date	Opp	AB	H	Hot or Cold?
1	4/5/1999	@Fla	4	2	H
2	4/6/1999	@Fla	4	1	C
3	4/7/1999	@Fla	4	1	C
4	4/8/1999	@Mon	2	0	C
5	4/9/1999	@Mon	3	0	C
6	4/10/1999	@Mon	5	3	H
7	4/11/1999	@Mon	5	4	H
8	4/12/1999	Fla	3	0	C
9	4/14/1999	Fla	3	0	C
10	4/15/1999	Fla	2	0	C
11	4/16/1999	Mon	4	2	H
12	4/17/1999	Mon	3	1	H
13	4/18/1999	Mon	2	2	H

then he will have long runs of hot games and long runs of cold games, so the number of runs in the sequence will be small.

Here we have discussed ways of detecting the observed streakiness that we see in baseball data. However, that does not mean that the player (or team) is truly streaky. In the next case study, we'll make a clear distinction between a player's streaky ability (or lack of streaky ability) and his streaky performance during a baseball season. We will see that a fair die, an object with consistent ability, can exhibit very streaky performance.

Case Study 8-6: A Streaky Die

Topics Covered: Moving averages, runs, simulation of coin tossing, patterns in coin tossing data   In the previous case study, we looked for streaky behavior in the day-to-day hitting data for John Olerud in the 1999 baseball season. Specifically, we looked at

- moving averages using a window of nine games—we were looking for unusually high or low moving averages to say that Olerud is streaky,

- runs of good and bad hitting days—here we looked at the total number of runs, and the length of the longest run; a small total number of runs and/or a long run of (hot or cold) games would indicate streakiness.

Using moving averages and runs, we saw some interesting features in Olerud's hitting data:

- In one nine-game stretch, Olerud hit .500; in another nine-game stretch, he only hit .167.

- Olerud had a total of 79 runs (either runs of Hot or runs of Cold). His longest run had length 5.

But are these really interesting? Do they mean that Olerud is truly a streaky hitter? In other words, can we tell from these data that Olerud has an ability to be streaky? Maybe he isn't really streaky and by luck or chance variation, we just happened to see some interesting streaky behavior.

Mr. Consistent

Let's consider this last question more carefully. Suppose that Olerud is truly not streaky. What would that mean?

The opposite of a streaky hitter is a consistent hitter. This type of hitter always gets a hit with the same probability no matter what the time or situation. This consistent hitter will get a hit with the same probability p in every at-bat in the season.

We can represent hitting of a consistent hitter by use of a die. Consider a 10-sided die with the sides 0, 1, 2, 3, 4, 5, 6, 7, 8, 9. Suppose we roll the die and a "hit" is recorded if the die rolls 1, 2, 3; otherwise, the hitter is out. Then the probability of a hit will be $\frac{3}{10} = .3$. More importantly, the chance of a hit is always .3 and the outcome of the die won't depend on what rolls occurred in the past.

We can use this die to simulate John Olerud's hit outcomes if he were really a consistent hitter. We will simulate hitting data for all 162 games by working in groups. Group 1 will simulate Olerud's hitting for games 1–18, group 2 will simulate Olerud's hitting for games 19–36, etc. When we complete this activity we will see if this simulated data looks streaky. In particular, we will look at

- moving averages using a window width of nine games,

- the number of runs and the length of the longest run.

What I think we'll discover is that this simulated data can look pretty streaky. Even when a hitter is truly consistent, there can be interesting patterns in the moving average graph that can be interpreted as streaky behavior. Figure 8.12 shows the moving average graph (using a width of nine games) for the data that a statistics class generated using a 10-sided die.

We see a lot of interesting patterns. Our hitter had an early slump around game 20 and had one significant hot streak where he batted in the .500 range. But remember this hitter is truly a consistent batter—what we are observing is the streakiness that is inherent in chance variation.

To show you that this is not a fluke occurrence, I did our simulation five times on a computer. I'm using the same assumptions as above.

1. Our player is a truly consistent hitter with probability $p = .3$ of getting a hit on a single at-bat.

2. I use the actual at-bat numbers for Olerud for the season of 162 games.

Figure 8.13 shows the moving average plots for my five simulations (I'm still using a window width of nine games).

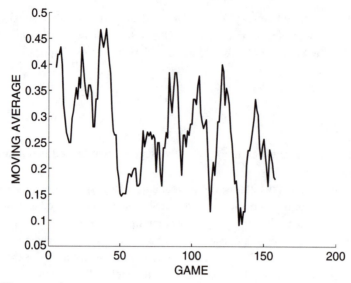

FIGURE 8.12
Moving average plot of simulated hitting data for a truly consistent hitter where the probability of a hit is equal to .3.

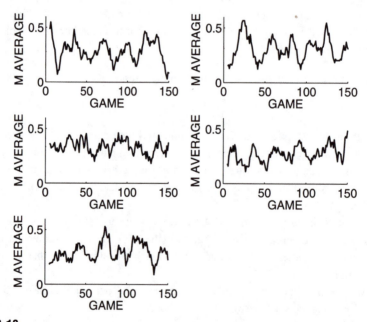

FIGURE 8.13
Moving average plots for five simulated datasets for a consistent hitter with a constant probability of .3 of getting a hit.

You should notice a lot of up-and-down behavior in the five moving average plots. So even if Olerud were a truly consistent hitter, he would likely have several interesting hot and cold streaks.

The moral of the story is that you should be cautious about interpreting streaky behavior of baseball players. Dice have a clearly consistent ability—just the opposite from true streakiness. But we've seen that dice can look streaky!

Exercises

Leadoff Exercise. Table 8.4 gives some situational OBPs for Rickey Henderson in the 1999 baseball season. This table shows how Rickey did for day and night games, for games home and away, for games played on grass and turf fields, for games played in domed and open ballparks, and against left- and right-handed pitchers.

TABLE 8.4
Situational on-base percentages of Rickey Henderson for the 1990 season.

Breakdown	AB	BB	PA	OBP
Day	144	21		.400
Night	294	61		.433
Home	200	34		.376
Away	238	48		.462
Grass	364	60		.408
Turf	74	22		.490
Dome	43	10		.472
Open	395	72		.417
vs. Right	333	54		.403
vs. Left	105	28		.481

(a) For each breakdown, compute the number of (approximate) plate appearances (PA) by adding at-bats (AB) to walks (BB).

(b) Let p_{day} denote Rickey's on-base probability when he is playing a day game and let p_{night} denote his on-base probability when he's playing at night. Using the data in the table, construct 90% probability intervals for p_{day} and p_{night}. Comparing the two intervals, can you conclude that Rickey really has a higher on-base probability at night games?

(c) Using the same method as in (b), compare the home/away OBPs, the grass/turf OBPs, the dome/open OBPs, and the right/left OBPs.

(d) Based on your work in (c) and (d), rank the five situations with respect to the most significant effect to the least significant effect.

8.1. Table 8.5 gives batting statistics for 20 randomly selected players in the 1998 season. The first three columns give the number of at-bats (AB), hits (H), and batting average (AVG) for all games played at home, and the next three columns give the same statistics for all games played away from home.

TABLE 8.5
Home and away batting statistics for 20 randomly selected players in the 1998 season.

Name	Home			Away			Home — Away
	AB	H	AVG	AB	H	AVG	
Desi Relaford	246	70	.285	248	51	.206	
Tony Clark	296	86	.291	306	89	.291	
Jose Cruz Jr.	188	45	.239	164	44	.268	
Bret Boone	293	77	.263	290	78	.269	
Rico Brogna	275	77	.280	290	73	.252	
Chris Widger	214	51	.238	203	46	.227	
Lenny Webster	165	55	.333	144	33	.229	
Fernando Tatis	287	84	.293	245	63	.257	
John Mabry	174	44	.253	203	50	.246	
Travis Lee	286	81	.283	276	70	.254	
Matt Williams	262	73	.279	248	63	.254	
Charlie Hayes	165	51	.309	164	43	.262	
Kevin Orie	168	31	.185	211	52	.246	
Ricky Gutierrez	239	60	.251	252	68	.270	
Lee Stevens	168	57	.339	176	34	.193	
Stan Javier	197	65	.330	220	56	.255	
Fernando Vina	314	85	.271	323	113	.350	
Brad Ausmus	199	61	.307	213	50	.235	
Bill Spiers	191	50	.262	193	55	.285	
Pat Meares	261	76	.291	282	65	.230	

(a) For each player, compute the difference in batting averages

$$DIFF = AVG\ (home) - AVG\ (away)$$

and put the differences in the Home — Away column of Table 8.5.

(b) Draw a stemplot of the batting average differences that you computed in part (a).

(c) Looking at the stemplot constructed in part (b), what is the median value of the difference? Also find the smallest and largest differences, and find the players that had these extreme values.

(d) From your work, can you say that players generally hit better at home? If so, by how much on the average?

8.2. Table 8.6 gives 1998 batting statistics for 20 players before the All-Star break and after the All-Star break.

TABLE 8.6
Batting statistics for 20 players before and after the 1998 All-Star break.

Name	Pre All-Star			Post All-Star			Pre − Post
	AB	H	AVG	AB	H	AVG	
Desi Relaford	252	75	.298	242	46	.190	
Tony Clark	309	85	.275	293	90	.307	
Jose Cruz Jr.	159	34	.214	193	55	.285	
Bret Boone	337	96	.285	246	59	.240	
Rico Brogna	301	81	.269	264	69	.261	
Chris Widger	270	61	.226	147	36	.245	
Lenny Webster	163	42	.258	146	46	.315	
Fernando Tatis	268	73	.272	264	74	.280	
John Mabry	213	56	.263	164	38	.232	
Travis Lee	338	96	.284	224	55	.246	
Matt Williams	317	86	.271	193	50	.259	
Charlie Hayes	189	57	.302	140	37	.264	
Kevin Orie	173	30	.173	206	53	.257	
Ricky Gutierrez	244	71	.291	247	57	.231	
Lee Stevens	220	54	.245	124	37	.298	
Stan Javier	280	77	.275	137	44	.321	
Fernando Vina	341	104	.305	296	94	.318	
Brad Ausmus	219	55	.251	193	56	.290	
Bill Spiers	213	62	.291	171	43	.251	
Pat Meares	296	75	.253	247	66	.267	

(a) Repeat parts (a)–(c) of Exercise 8.1 for this dataset. Here the Pre − Post column of Table 8.6 will contain the differences in batting average

$$\text{AVG(Pre All-Star Game)} - \text{AVG(Post All-Star Game)}.$$

(b) From your work, can you say that players generally play better after or before the All-Star break? If so, how much on the average?

8.3. Table 8.7 displays 1998 batting statistics for 20 players when the batter was ahead in the count and when the batter had two strikes in the count.

(a) Repeat parts (a)–(c) of Exercise 8.1 for this dataset. Here the Ahead − 2 Strikes column of Table 8.7 will contain the differences in batting average

$$\text{AVG(Ahead in Count Pitch Count)} - \text{AVG(2 Strikes Pitch Count)}.$$

TABLE 8.7

Batting statistics for 20 players ahead in the count and having two strikes.

Name	Ahead in Count			Two Strikes			Ahead − 2 Strikes
	AB	H	AVG	AB	H	AVG	
Desi Relaford	87	32	.368	230	39	.170	
Tony Clark	120	53	.442	292	53	.182	
Jose Cruz Jr.	76	20	.263	161	23	.143	
Bret Boone	132	44	.333	291	64	.220	
Rico Brogna	116	36	.310	274	56	.204	
Chris Widger	77	21	.273	212	46	.217	
Lenny Webster	84	28	.333	119	26	.218	
Fernando Tatis	102	39	.382	270	54	.200	
John Mabry	74	19	.257	173	26	.150	
Travis Lee	108	33	.306	257	43	.167	
Matt Williams	110	39	.355	203	33	.163	
Charlie Hayes	78	30	.385	151	27	.179	
Kevin Orie	79	19	.241	159	27	.170	
Ricky Gutierrez	76	22	.289	229	48	.210	
Lee Stevens	70	20	.286	160	21	.131	
Stan Javier	92	36	.391	175	30	.171	
Fernando Vina	158	59	.373	191	42	.220	
Brad Ausmus	95	37	.389	161	27	.168	
Bill Spiers	86	34	.395	166	33	.199	
Pat Meares	116	39	.336	240	45	.188	

(b) From your work, can you say that players generally hit better when they are ahead in the count as opposed to two strikes? If so, how much on the average?

8.4. Suppose that we are interested in how players hit on odd-numbered days (like April 7, May 13, June 9) as opposed to even-numbered days (like April 10, May 14, July 20). Here we are pretty sure that there is no true situational effect for any player. (Why would any player actually be a better or worse hitter on odd-numbered days?) We simulate the type of hitting data that one might see in this scenario by means of the following experiment. For each of 20 players, we first simulate their abilities (their hitting probabilities) using a normal curve with mean .276 and standard deviation 0.021. Here are the twenty simulated abilities:

.267	.241	.279	.282	.252	.301	.301	.275	.283	.280
.273	.278	.298	.277	.274	.259	.272	.291	.264	.322

Then we simulate the situational data as follows. For each player, we simulate the results of 300 at-bats for the odd-numbered days using the above hitting probabilities, and then simulate the results of 300 at-bats for the even-numbered days using the same set of hitting probabilities. We obtain the data in Table 8.8.

TABLE 8.8
Simulated batting statistics for 20 players on even-numbered and odd-numbered days.

Name	Even-Numbered Days			Odd-Numbered Days			Even − Odd
	AB	H	AVG	AB	H	AVG	
Jim	300	74	.247	300	74	.247	
Pat	300	74	.247	300	80	.267	
Edsel	300	76	.253	300	92	.307	
Arjun	300	72	.240	300	85	.283	
Rich	300	70	.233	300	75	.250	
Pete	300	105	.350	300	83	.277	
Sam	300	80	.267	300	89	.297	
Dave	300	73	.243	300	84	.280	
Curt	300	66	.220	300	91	.303	
Ben	300	80	.267	300	96	.320	
Adam	300	104	.347	300	77	.257	
Dale	300	91	.303	300	79	.263	
Joe	300	79	.263	300	65	.217	
Dick	300	95	.317	300	99	.330	
Brad	300	88	.293	300	85	.283	
Hal	300	83	.277	300	74	.247	
Jay	300	77	.257	300	91	.303	
John	300	80	.267	300	67	.223	
Charles	300	74	.247	300	87	.290	
Neal	300	81	.270	300	81	.270	

(a) Compute all of the batting average differences

$$\text{DIFF} = \text{AVG(Even-Numbered Days)} - \text{AVG(Odd-Numbered Days)}$$

and put the differences in the Even − Odd column of Table 8.8.

(b) Graph the batting average differences using a stemplot.

(c) Find the average difference, the low and high differences, and find the batters who had these extreme values.

(d) Do these simulated data resemble any of the situational data described in the chapter? Explain.

8.5. Suppose that it is found out that some ballparks are using a special baseball that travels further when hit. Moreover, it is known that this special baseball will add

TABLE 8.9

Simulated hitting data for 20 players using a special baseball and the usual baseball when there is a true situational bias.

Name	Special Baseball			Usual Baseball			Special − Usual
	AB	H	AVG	AB	H	AVG	
Larry	300	98	.327	300	103	.343	
Moe	300	84	.280	300	68	.227	
Curly	300	102	.340	300	79	.263	
Harold	300	102	.340	300	88	.293	
Harvey	300	83	.277	300	94	.313	
Lee	300	90	.300	300	91	.303	
Charles	300	98	.327	300	85	.283	
Bill	300	74	.247	300	68	.227	
Bob	300	92	.307	300	69	.230	
Rick	300	95	.317	300	85	.283	
Burt	300	97	.323	300	92	.307	
Lynn	300	89	.297	300	94	.313	
Marvin	300	92	.307	300	91	.303	
Randy	300	94	.313	300	88	.293	
Tom	300	98	.327	300	84	.280	
Joseph	300	99	.330	300	82	.273	
Jimmy	300	116	.387	300	107	.357	
Bret	300	87	.290	300	74	.247	
Britt	300	87	.290	300	76	.253	
Clark	300	94	.313	300	75	.250	

30 points to every player's hitting probability. Table 8.9 generates some simulated hitting data using this scenario.

To simulate these data, we first simulate hitting probabilities for the 20 players using a normal curve with mean .261 and standard deviation 0.021. These probabilities correspond to the abilities of the hitters playing with the usual ball.

.282	.248	.291	.310	.261	.294	.302	.243	.246	.288
.303	.290	.301	.251	.276	.273	.268	.290	.293	.291

To obtain the hitting probabilities for the players with the special ball, we add 30 points to each of the usual hitting probabilities.

.312	.278	.321	.340	.291	.324	.332	.273	.276	.318
.333	.320	.331	.281	.306	.303	.298	.320	.323	.321

Then we simulate hitting data for 300 at-bats using the usual probabilities and for 300 at-bats using the special probabilities—we show the data in Table 8.9.

(a) Compute all of the batting average differences

$$DIFF = AVG(\text{Special Ball}) - AVG(\text{Usual Ball})$$

and put the differences in the Special − Usual column of Table 8.9.

(b) Graph the batting average differences using a stemplot.

(c) Find the average difference, the low and high differences, and find the batters who had these extreme values.

(d) Do these simulated data resemble any of the situational data described in the chapter? Explain.

8.6. (Exercise 8.5 continued)   Suppose that players have different abilities to use the special ball that was discussed in Exercise 8.5. For the first seven players listed in Table 8.10 (the first group), the special ball adds 50 points to the Usual ball hitting probability. For the next six players in the table (the second group), the special ball adds 30 points to the player's hitting probability, and for the final seven players (the

TABLE 8.10

Simulated hitting data for 20 players using a special baseball and the usual baseball when there are ability situational effects.

		Special Baseball			Usual Baseball			
	Name	AB	H	AVG	AB	H	AVG	Special − Usual
First group	Larry	300	104	.347	300	98	.327	
	Moe	300	86	.287	300	65	.217	
	Curly	300	99	.330	300	87	.290	
	Harold	300	116	.387	300	102	.340	
	Harvey	300	83	.277	300	69	.230	
	Lee	300	116	.387	300	93	.310	
	Charles	300	97	.323	300	90	.300	
Second group	Bill	300	81	.270	300	66	.220	
	Bob	300	82	.273	300	62	.207	
	Rick	300	89	.297	300	80	.267	
	Burt	300	91	.303	300	85	.283	
	Lynn	300	94	.313	300	81	.270	
	Marvin	300	106	.353	300	95	.317	
Third group	Randy	300	65	.217	300	73	.243	
	Tom	300	75	.250	300	89	.297	
	Joseph	300	84	.280	300	82	.273	
	Jimmy	300	75	.250	300	74	.247	
	Bret	300	85	.283	300	86	.287	
	Britt	300	86	.287	300	82	.273	
	Clark	300	96	.320	300	87	.290	

third group), the special ball only adds ten points to the hitting probability. Below we give the hitting probabilities for the 20 players.

Players	Hitting Probabilities for Usual Ball		Hitting Probabilities for Special ball
First group	.282 .248 .291 .310 .261 .294 .302	Add 50 points	.332 .298 .341 .360 .311 .344 .352
Second group	.243 .246 .288 .303 .290 .301	Add 30 points	.273 .276 .318 .333 .320 .331
Third group	.251 .276 .273 .268 .290 .293 .291	Add 10 points	.261 .286 .283 .278 .300 .303 .301

Table 8.10 shows hitting data simulated from this model.

(a) Compute the differences in batting average (Special Ball − Usual Ball) and put the differences in the last column of the table.

(b) Construct a stemplot of the differences.

(c) Compute the average batting average difference, and find the smallest and largest differences.

(d) Compare the batting average differences with the differences found in Exercise 8.5. Can you offer any explanation for the different patterns that you see in stemplots from Exercises 8.5 and this exercise?

8.7. In Exercise 8.1, we looked at home and away batting averages for twenty players during the 1998 season. Table 8.11 gives home and away averages for the same twenty players, but for the 1999 season.

(a) Compute the differences in batting averages AVG(home) − AVG(away) and place your work in Table 8.11.

(b) Construct a stemplot of the batting average differences. Based on this graph, would you say that there is a general tendency for players to bat better at home? Why or why not?

(c) Suppose that some batters hit unusually well at home, and other batters actually hit better on the road. If players did have different abilities to bat at home games versus away games, then one would expect to see a relationship between the batting average difference (Home − Away) for 1998 and the batting average difference for 1999. Using the data from Tables 8.5 and 8.11, construct a scatterplot of the 1998 and 1999 differences.

(d) Based on your scatterplot drawn in (c), do you see a relationship between a player's batting average difference for 1998 and 1999? Interpret what this means in terms of one's ability to hit well at home versus away games.

TABLE 8.11
Home and away batting averages for twenty players in the 1999 season.

	1999		
Name	Home	Away	Home − Away
Desi Relaford	.245	.239	
Tony Clark	.261	.296	
Jose Cruz Jr.	.267	.210	
Bret Boone	.271	.234	
Rico Brogna	.303	.255	
Chris Widger	.294	.229	
Lenny Webster	.148	.087	
Fernando Tatis	.276	.321	
John Mabry	.223	.265	
Travis Lee	.301	.181	
Matt Williams	.279	.326	
Charlie Hayes	.202	.207	
Kevin Orie	.262	.246	
Ricky Gutierrez	.303	.214	
Lee Stevens	.275	.290	
Stan Javier	.269	.298	
Fernando Vina	.230	.300	
Brad Ausmus	.291	.259	
Bill Spiers	.260	.315	
Pat Meares	.152	.397	

8.8. Situational data are also recorded for pitchers. Table 8.12 displays pitching statistics for 20 pitchers for the three-year period 1997–1999. For each pitcher, the table gives

Throws: the throwing hand of the pitcher,

Batting Average—Left: the batting average of left-handed hitters against the pitcher,

Batting Average—Right: the batting average of right-handed hitters against the pitcher,

Batting Average—Pitch 1–15: the batting average of the hitters on pitches 1–15 during a game,

Batting Average—Pitch 46–60: the batting average of the hitters on pitches 46–60 during a game,

Batting Average—Pitch 91–105: the batting average of the hitters on pitches 91–105 during a game.

(a) If a pitcher is right-handed, then who will be a more effective hitter—a left-hander or a right-hander? Why?

TABLE 8.12

Situational statistics for 20 pitchers for the three-year period 1997–1999.

Pitcher	Throws	Batting Average Left	Batting Average Right	DIFF1 = OPP − SAME	Batting Average– Pitch 1–15	46–60	91–105	DIFF2
John Smoltz	Right	.252	.227		.293	.216	.260	
Alan Ashby	Right	.271	.250		.266	.279	.284	
Alex Fernandez	Right	.284	.203		.283	.212	.208	
Andy Pettite	Right	.295	.266		.255	.287	.302	
Bartolo Colon	Right	.267	.248		.248	.282	.219	
Chuck Finley	Left	.236	.249		.226	.294	.254	
Charles Nagy	Right	.291	.291		.294	.298	.272	
David Cone	Right	.237	.219		.212	.260	.273	
Doug Drabek	Left	.282	.292		.260	.261	.254	
Darryl Kile	Right	.284	.251		.299	.296	.267	
Greg Maddux	Right	.244	.254		.284	.245	.216	
Kevin Brown	Right	.249	.216		.259	.264	.221	
Randy Johnson	Left	.178	.213		.216	.221	.258	
Curt Schilling	Right	.244	.220		.250	.210	.232	
Tom Glavine	Left	.250	.251		.283	.256	.224	
Terry Mulholland	Left	.273	.270		.253	.269	.266	
Pedro Martinez	Right	.210	.193		.171	.204	.221	
Todd Stottlemyre	Right	.279	.216		.259	.242	.300	
Wilson Alvarez	Right	.258	.238		.291	.244	.224	
Mike Mussina	Left	.244	.251		.257	.246	.238	

(b) If a pitcher is left-handed, then who will be a more effective hitter—a left-hander or a right-hander?

(c) Define an opposite-side hitter as one who bats at the opposite side from the arm of the pitcher. Likewise, define a same-side hitter as a hitter who bats from the same side as the arm of the pitcher. For each pitcher, compute the difference

$$\text{DIFF1} = \text{AVG(opposite-side hitter)} - \text{AVG(same-side hitter)}.$$

Put your batting average differences in the DIFF1 column of Table 8.12.

(d) Graph the differences and find the average value. Conclude what you have learned about the effectiveness of opposite-side hitters compared to same-side hitters.

8.9. (Exercise 8.8 continued) Table 8.12 also shows the batting average of hitters during different pitch counts of each pitcher.

(a) How do you expect a starting pitcher to perform at the beginning of a game? Do you think pitchers need to warm up and don't reach full effectiveness until they have pitched a few innings?

(b) How do you expect a starting pitcher to perform at the end of the game after he has thrown 90 pitches? Is fatigue a factor for a starting pitcher?

(c) If a starting pitcher gets tired, what effect would that have on the batting average of opposing hitters?

(d) To see how a pitcher performs in the middle of the game as opposed to the beginning, we can compute the difference

$$\text{DIFF2} = \text{AVG(pitches 46–60)} - \text{AVG(pitchers 1–15)}.$$

For each pitcher, compute this difference and put the values in the DIFF2 column of Table 8.12.

(e) Graph the differences you computed in part (d) and compute an average value. Can you detect a general tendency for pitchers to tire out between the beginning and middle parts of the game?

8.10. (Exercise 8.8 continued) If we thought that pitchers generally tire out during a game, then we would expect the opposing batting average to be larger for pitch count 91–105 than for pitch count 46–50.

(a) In Table 8.12, count the number of pitchers who pitch better on pitches 91–105 (compared to pitches 46–50), and count the number of pitchers who do better on pitches 46–60. Put the counts in the table below. Next, find the proportion in each group and put the values in the Proportion column.

	Count	Proportion
Pitchers who do better on pitches 91–105		
Pitchers who do better on pitches 46–60		

(b) What have you learned from the proportions you computed in (a)?

(c) Suppose that you now look at how batters perform during different pitch counts. If batters generally have a smaller AVG for pitches 91–105, does that mean that pitchers tend to get stronger during a baseball game? Explain why this might be a wrong conclusion.

8.11. (Exercise 5.1 continued) Table 8.13 shows Doug Glanville's daily batting record for the first 20 games in the 1999 baseball season. To see how Glanville performs in short periods, one can compute moving averages of five games. For example, in games 1 through 5, Table 8.13 shows that Glanville had four hits in 17 at-bats for a batting average of $\frac{4}{17} = .235$. In the next group of five games (2 through 6), Glanville was 3 for 18 for a batting average of $\frac{3}{18} = .167$.

(a) For each group of five games, find the number of at-bats, hits, and five-game moving batting average. Put the answers in the table.

TABLE 8.13
Computation of moving averages for Doug Glanville's 1999 daily batting record.

Game	AB	H	5 Games	5-Game AB	5-Game H	Moving AVG
1	3	1				
2	4	1				
3	4	1	1 through 5	17	4	$\frac{4}{17} = .235$
4	3	1	2 through 6	18	3	$\frac{3}{18} = .167$
5	3	0	3 through 7			
6	4	0	4 through 8			
7	5	2	5 through 9			
8	5	1	6 through 10			
9	6	3	7 through 11			
10	5	1	8 through 12			
11	5	3	9 through 13			
12	4	0	10 through 14			
13	4	3	11 through 15			
14	3	1	12 through 16			
15	3	1	13 through 17			
16	4	1	14 through 18			
17	4	1	15 through 19			
18	5	3	16 through 20			
19	4	1				
20	4	0				

(b) Graph the moving averages against the middle-game number on the grid below. (For example, you would graph the first moving average .235 against the mid-

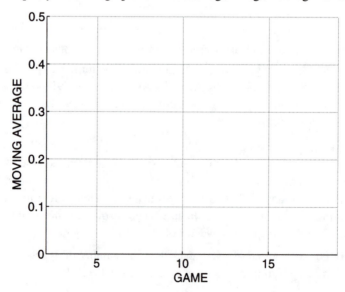

game number 3, the moving average .167 against the mid-game number 4, and
so on.)

(c) Describe the pattern that you see in the moving average plot. Are there any periods that Glanville appears to be hot or cold?

8.12. (Exercise 8.11 continued) Table 8.14 shows Glanville's day-to-day batting performance for the first 20 games of the 1999 season. Glanville's batting average for the
whole season was .300. We say that Glanville is hot for a particular game if his game
batting average is over .300; otherwise we say that Glanville is cold. For example, on
game 1, his game average was $\frac{1}{3} = .333$. Since this is larger than .300, we say that
he was hot in game 1.

TABLE 8.14
Doug Glanville's 1999 daily batting record for the first 20 games.

Game	AB	H	Game AVG	Hot or Cold
1	3	1	.333	Hot
2	4	1		
3	4	1		
4	3	1		
5	3	0		
6	4	0		
7	5	2		
8	5	1		
9	6	3		
10	5	1		
11	5	3		
12	4	0		
13	4	3		
14	3	1		
15	3	1		
16	4	1		
17	4	1		
18	5	3		
19	4	1		
20	4	0		

(a) Complete Table 8.14. For each game, compute the game batting average and
classify the game as hot or cold.

(b) From the sequence of Hot and Cold games, find

  i. the longest run of Hot games,

  ii. the longest run of Cold games,

  iii. the total number of runs.

8.13. (Exercise 8.11 continued) For the game-to-game batting data for Doug Glanville, count the number of games that Glanville had 0 hits, 1 hit, etc. Put your counts in the table below.

Number of hits	0	1	2	3
Count				

8.14. Table 8.15 shows the game results for the Philadelphia Phillies and the Atlanta Braves for the first thirty games of the 2000 season.

TABLE 8.15

Game results for Philadelphia Phillies and Atlanta Braves for first thirty games of the 2000 season.

Phillies

L L L W L W W W L L W L L L W L L L L L W L L L L W W L W W

Braves

W L W L W W L L L W L W W W W W W W W W W W W W W L W L L

(a) Compute all moving fractions of wins using a width of 9 games for both teams. That is, find the proportion of wins in games 1-9, in games 2–10, . . . , in games 22–30. Put your results in the table below.

		Phillies	Braves			Phillies	Braves
Games	Mid game	Fraction of wins	Fraction of wins	Games	Mid game	Fraction of wins	Fraction of wins
1–9	5			12–20	16		
2–10	6			13–20	17		
3–11	7			14–22	18		
4–12	8			15–23	19		
5–13	9			16–24	20		
6–14	10			17–25	21		
7–15	11			18–26	22		
8–16	12			19–27	23		
9–17	13			20–28	24		
10–18	14			21–29	25		
11–19	15			22–30	26		

(b) Graph the moving fractions against the mid-game for the two teams on the axes below.

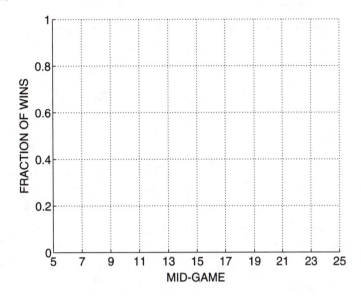

(c) Comment on the pattern of moving fractions that you see in the graph. Did the Phillies or Braves seem unusually hot or cold during these first 30 games?

8.15. (Exercise 8.14 continued)   For each of the win/loss sequences of the Phillies and Braves shown, find (1) the longest run of wins or losses, and (2) the total number of runs. Put your answers in the table below.

Team	Longest run of wins or losses	Number of runs
Atlanta		
Philadelphia		

8.16. (Exercise 8.12 continued)   Suppose Doug Glanville is truly a consistent hitter. The chance that he gets a hit on a single at-bat is .3 (his true batting average), and the results of different at-bats are independent. Suppose we simulate data from this model using Glanville's actual number of at-bats for the 20 games. The results of this simulation are placed in Table 8.16.

(a) As in Exercise 8.12, classify the hitting of each game as either hot or cold, depending if the game batting average is larger or smaller than .3. Put your results in the table.

(b) Compute

   i. the longest run of Hot games,

   ii. the longest run of Cold games,

   iii. the total number of runs.

TABLE 8.16

Simulated data for 20 games assuming Glanville is a consistent hitter with hit probability of .3.

Game	AB	H	Game AVG	Hot or Cold
1	3	1	.333	Hot
2	4	2		
3	4	0		
4	3	1		
5	3	1		
6	4	1		
7	5	2		
8	5	3		
9	6	1		
10	5	2		
11	5	3		
12	4	0		
13	4	0		
14	3	0		
15	3	2		
16	4	0		
17	4	0		
18	5	0		
19	4	1		
20	4	0		

(c) Compare the results of these simulated data (the longest run of Hot, the longest run of Cold, and the total number of runs) with the results using Glanville's actual data in Table 8.14. Is there any evidence that Glanville is a streaky hitter?

8.17. (Exercise 8.16 continued)   In Exercise 8.16, we assumed that Glanville was really a consistent hitter with probability of .3 of getting a hit, and simulated 20 games of hitting assuming this consistent model. Suppose we repeat this experiment 20 times. For each experiment, we simulate 20 games of hitting (assuming Glanville is really a consistent hitter), and then classify each game as either hot or cold. The results are shown in Table 8.17.

(a) For each experiment, find the longest run of either Hot or Cold. Put the longest run in the Longest run column of the table.

(b) Construct a dotplot of the longest run values using the number line below.

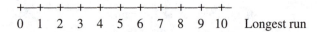

(c) For Glanville's data in Table 8.14, the longest run (of either Hot or Cold) was ____.

TABLE 8.17

Simulated data for twenty experiments, where one experiment consists of 20 games assuming Glanville is a consistent hitter with hit probability of .3. Glanville's performance in each game is classified as hot (H) or cold (C).

														Game						
1	2	3	4	5	6	7	8	9	10	11	12	13	14	15	16	17	18	19	20	Longest run
H	H	C	H	H	C	H	H	C	H	H	C	C	C	H	C	C	C	C	C	
C	H	C	H	H	H	C	H	C	C	H	H	H	C	H	C	C	H	H	C	
H	C	C	C	H	C	C	C	C	C	C	C	C	H	C	C	C	C	C	C	
H	C	C	H	C	H	C	C	H	C	H	H	C	C	C	C	C	C	C	C	
H	H	C	C	H	H	H	C	H	C	C	H	H	H	H	C	C	C	C	H	
H	H	C	H	H	H	H	H	H	C	C	C	C	C	H	H	H	H	C	H	
H	C	H	H	H	C	H	H	H	C	C	C	H	C	C	H	C	H	C	C	
H	C	H	H	C	H	H	H	H	H	H	H	C	C	H	H	H	H	C	C	
H	C	H	H	C	C	H	C	C	C	H	C	C	C	H	C	C	H	C	C	
H	H	C	H	C	H	C	C	C	H	H	C	C	H	C	C	C	C	C	H	
H	C	C	H	C	C	C	C	C	C	C	H	C	H	H	H	C	C	C	C	
H	H	H	H	C	C	H	C	C	H	C	C	C	C	H	H	C	H	H	H	
H	C	C	H	C	C	H	C	H	H	C	C	H	H	C	C	C	H	C	C	
H	C	C	H	H	H	C	H	C	C	H	C	H	H	H	C	H	H	C	H	
H	H	C	H	H	H	C	C	H	C	C	C	H	H	H	H	C	H	C	H	
H	H	C	C	H	C	H	H	C	C	H	C	C	H	H	C	H	C	C		
C	C	C	C	H	H	H	C	H	H	C	C	H	H	H	H	C	C	H	C	
H	C	C	C	H	C	C	H	C	C	C	C	C	C	H	C	C	C	C		
H	C	C	H	H	C	C	C	H	H	H	H	C	H	H	C	C	H	H	C	
C	C	C	C	H	C	C	H	C	C	H	C	H	H	H	H	C	C	C	H	

Does this value seem unusually small or large if Glanville was really a consistent hitter? (Compare Glanville's value with the longest run values plotted in part (b).)

8.18. Was Barry Bonds a streaky home run hitter in 2001? Using Bonds' home run log from Exercise 5.1, one can construct a moving average plot of his home run rate. The moving averages are shown in Figure 8.14 using a window of ten games. We see that Bonds showed some streaky behavior—his 10-game home run rate was over .25 about game 40, and his home run rate dropped to 0 about game 75.

How would Bonds perform if he were truly a consistent home run hitter? In 2001, Bonds hit 73 home runs in 664 plate appearances for a season rate of $73/664 = .1099$. Suppose that the probability that Bonds hits a home run in a single plate appearance is $p = .1099$, and the results of different plate appearances are independent. Using this consistent model, four seasons of home run hitting were simulated using the same game-to-game plate appearances as Bonds. Moving average plots of the home run rates for the four simulations are shown in Figure 8.15.

Discuss any unusual features of each of these four simulations. Do you think that Bonds' moving average plot is different, with respect to streakiness, from the plots of

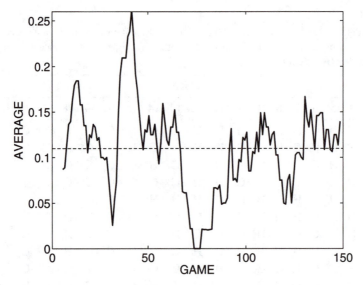

FIGURE 8.14
Moving average plot of 2001 Barry Bonds' home run rates using a window of ten games.

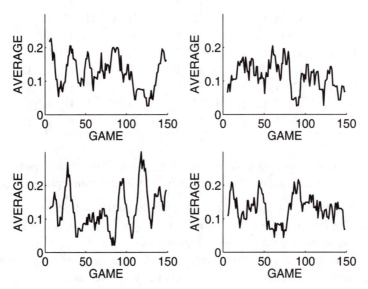

FIGURE 8.15
Moving average plot of home run rates using a consistent model with home run probability .1099.

the four simulated "consistent" hitters? Do you think that Bonds was a truly streaky home run hitter?

8.19. On the baseball-reference.com web site, one can explore the patterns of hot streaks and slumps of any Major League baseball team in history. (Look for the "streaks" link on the web site.) The Oakland Athletics had an interesting pattern of wins and losses during the 2002 season. Figure 8.16 displays a moving average of the winning proportion using a window of 20 games.

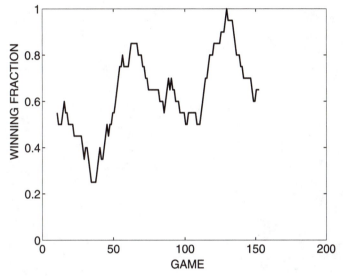

FIGURE 8.16
Moving average plot of winning fraction of the 2002 Oakland Athletics using a window of 20 games.

(a) Looking at Figure 8.16, describe the slumps and hot streaks of Oakland during this season.

(b) Oakland concluded this season with a 103–59 win/loss record with a winning percentage of 63.6%. Simulate sequences of wins and losses for an entire 162 game season assuming Oakland is a truly consistent team with probability of winning each game equal to .636. By comparing the simulated sequences with Figure 8.16, can you say that Oakland was truly a streaky team in the 2002 season?

(c) Choose another team in history that had a reputation for being unusually streaky during a particular season. By using the simulation method described in part (b), decide if the pattern of wins and losses for this team was different from that of a truly consistent team.

Further Reading

Situational baseball data for batters and pitchers is presented in *Stats Player Profiles 2002*. Chapter 4 of Albert and Bennett (2001) describes the different probability models that can be used for situational hitting data. Based on an analysis of situational data for the 1998 season, they classify the different true situational effects as "no effects," biases, or ability effects. Gilovich et al. (1985) describe the tendency of people to misinterpret the inherent streaky nature of random sequences. Albert and Bennett (2001), Chapter 5 discuss hitters who are genuinely consistent or streaky, and describe how one can perform inference about streakiness.

9

Modeling Baseball Using a Markov Chain

What's On-Deck?

In this chapter, we introduce a special probability model, called a Markov Chain, to represent the sequence of plays in a baseball game. The *state* of an inning is defined by the number of outs and the runners on the three bases. A half-inning of baseball can be regarded as a sequence of states until there are three outs. One can represent the random movement between states by a Markov Chain, where one moves from one state to another state with a given probability. In Case Study 9-1, we introduce a Markov Chain using an example of a person traveling between cities, and in Case Study 9-2 we extend the basic structure of a Markov Chain to a baseball game. To specify a Markov Chain, one needs only to specify the probabilities that control the movement between states, and we use actual baseball game data to estimate these probabilities. The remainder of the chapter shows the usefulness of this probability model to answer questions about baseball. Using the model, one can estimate the number of players that come to bat during the inning, and compute the probability that a team will score at least one run in an inning. One of the most useful computations is the expected number of runs scored from a particular state in an inning. Using these expected numbers of runs scored, one can assess the value of a particular hit, such as a home run. In addition, one can use these expected numbers of runs scored to judge the value of a particular batting play. Baseball fans are interested in the value of particular baseball strategies, such as a sacrifice bunt and a steal, and this probability model is useful in seeing if a particular strategy is helpful toward the general goal of scoring runs.

Case Study 9-1: Introduction to a Markov Chain

Topics Covered: Transition probabilities, absorbing states, expected number of visits to different states, matrices, matrix multiplication and inversion   In this chapter, we will show how a baseball game can be modeled using a special probability model called a

243

Markov Chain. In this first case study, we give a gentle introduction to Markov Chains and the remainder of the case studies will show the application of this idea to baseball.

Suppose a baseball fan is planning an interesting (you might call it bizarre) trip to New York City. He will start from San Francisco (SF). The next day, he will either stay in SF another night or he will fly to St. Louis (STL); he is equally likely to stay or fly to STL. If he does go to STL, then he will stay in STL another night with probability .4, or fly back to SF with probability .3, or fly to Chicago (CHI) with probability .3. If he is in CHI, the next day he is equally likely to stay another night, fly to SF, fly to STL, or fly to New York City (NY). Once he arrives at New York City, he will stay there—his trip is over.

Obviously the exact trip of this fan is random. All he knows for sure is that he will eventually arrive in New York City. But what is the probability that he will arrive in NY in exactly two days? In three days? How long will it take this fan, on average, to get to NY? And how many days can this fan expect to stay in SF, STL, and CHI?

The map in Figure 9.1 shows the possible city-to-city connections of our traveler. An arrow from one city to another city indicates the particular trip is possible and the number label is the probability of this trip. An arrow from a city back to the same city corresponds to a night where the traveler stays over.

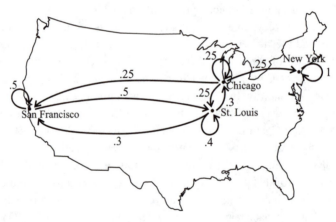

FIGURE 9.1
City-to-city connections of a traveler in trip from San Francisco to New York City.

A Markov Chain describes movement between a number of locations, called *states*. Here there are four states for our traveler, SF, STL, CHI, and NY. Given that the traveler is in a particular state (city), then we will move to another state the next day with specific probabilities. The probability that he will move to another city depends only on his current location and not on the previous cities he visited. (This is a special property of a Markov Chain.) Given the information above, we know if the person is currently in SF, then he will travel to

SF   STL   CHI   NY

with respective probabilities

$$.5 \quad .5 \quad 0 \quad 0$$

If he is currently in STL, then he travels to these four cities the next day with respective probabilities

$$.3 \quad .4 \quad .3 \quad 0$$

These probabilities are called *transition probabilities*—they describe the likelihoods of moving between various states in the Markov Chain. We summarize all of these transition probabilities by means of a transition matrix P shown in Table 9.1.

TABLE 9.1
Matrix of transition probabilities for the traveler example.

		SF	STL	CHI	NY
	SF	0.5	0.5	0	0
$P =$	STL	0.3	0.4	0.3	0
	CHI	0.25	0.25	0.25	0.25
	NY	0	0	0	1

The first row of this matrix P gives the transition probabilities of the traveler starting from SF, the second row gives the probabilities starting from STL, and so on. One special feature of this particular Markov Chain is that once the traveler arrives at New York City, he will remain there. We call the state NY an *absorbing state*—as indicated in the transition matrix, the probability of remaining in an absorbing state is one.

There are a number of nice results about Markov Chains that will make it easy to answer the questions posed above.

What Is the Probability of Reaching States After a Specific Number of Moves?

The matrix P gives the probabilities of reaching various states in exactly one move. We can obtain the probabilities of reaching states in two moves by squaring the matrix P:

$$P^2 = P \times P = \begin{bmatrix} .5 & .5 & 0 & 0 \\ .3 & .4 & .3 & 0 \\ .25 & .25 & .25 & .25 \\ 0 & 0 & 0 & 1 \end{bmatrix} \times \begin{bmatrix} .5 & .5 & 0 & 0 \\ .3 & .4 & .3 & 0 \\ .25 & .25 & .25 & .25 \\ 0 & 0 & 0 & 1 \end{bmatrix}$$

$$= \begin{bmatrix} .4 & .45 & .15 & 0 \\ .345 & .385 & .195 & .075 \\ .2625 & .2875 & .1375 & .3125 \\ 0 & 0 & 0 & 1 \end{bmatrix}$$

The first row of this matrix, [.4, .45, .15, 0], gives the probability that, starting at SF, we will be in the respective cities SF, STL, CHI, NY in exactly two days. So he will be in Chicago in two days with probability .15. Note that the probability that he'll be in NY in two days is 0.

If we multiply the matrix P additional times, we obtain the probabilities of being in the states after more than two days. If we multiply

$$P^6 = P \times P \times P \times P \times P \times P$$

we obtain the six-step transition probabilities

$$P^6 = \begin{bmatrix} .323 & .361 & .154 & .163 \\ .293 & .328 & .140 & .240 \\ .219 & .244 & .104 & .433 \\ 0 & 0 & 0 & 1 \end{bmatrix}$$

The first row of P^6 gives the probabilities of being in the four states, starting in SF, in exactly six days. We see that if we start in SF, the chance that we will be in NY in six days is .163.

How Long Will the Traveler Stay in the States?

If we remove the row and column corresponding to the single absorbing state from the matrix P, we have the matrix Q of transition probabilities

$$Q = \begin{bmatrix} 0.5 & 0.5 & 0 \\ 0.3 & 0.4 & 0.3 \\ 0.25 & 0.25 & 0.25 \end{bmatrix}$$

To find the expected number of times that the process visits the states before being absorbed, we simply compute the matrix $E = (I - Q)^{-1}$, where I is the 3 by 3 identity matrix.[1] Here

$$E = \left(I - \begin{bmatrix} 0.5 & 0.5 & 0 \\ 0.3 & 0.4 & 0.3 \\ 0.25 & 0.25 & 0.25 \end{bmatrix} \right)^{-1} = \begin{bmatrix} 10 & 10 & 4 \\ 8 & 10 & 4 \\ 6 & 6.67 & 4 \end{bmatrix}$$

Let's interpret this matrix. The first row, [10   10   4], gives the expected number of visits to the cities SF, STL, and CHI if we start at San Francisco. So in the trip, the person will be in SF, on average, ten days (including the starting day), STL for ten days, and CHI for four days. Likewise, the second and third rows of the matrix E give the expected number of visits if we start in St Louis and Chicago, respectively.

The matrix E tells us how the traveler will do on average. But actually there is a lot of variation in the length of the trip. To see this, I had the computer simulate 1000 trips origi-

---

[1]The expected number of visits matrix can be found by adding the matrices $Q, Q^2, Q^3, \ldots$. Then, for matrices of transition probabilities, it can be shown that $Q + Q^2 + Q^3 + \cdots = (I - Q)^{-1}$.

nating from San Francisco using the matrix of transition probabilities. Figure 9.2 shows a histogram of the lengths (in days) of the 1000 trips. The distribution is right-skewed where the length ranged from three to 142 days. The mean length of a trip was 23.58 days. This is reasonable—looking at the first row of the matrix E, we see that we will spend a total of $10 + 10 + 4 = 24$ days in our journey.

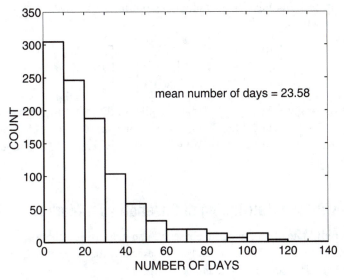

FIGURE 9.2
Histogram of lengths of 1000 trips from San Francisco simulated from the Markov Chain.

How Many Frequent Flyer Miles?

Suppose our traveler plans on flying and he is interested in how many frequent flyer miles he will accumulate in his trip. First we record in a table the distances between all of these cities, as displayed in Table 9.2.

TABLE 9.2
Distances between all cities in the traveler example.

	SF	STL	CHI	NY
SF	0	2062	2133	2908
$P =$ STL	2062	0	294	953
CHI	2133	294	0	790
NY	2908	953	790	0

Then we find the expected length of a one-day trip from each city. If we start in SF, then we will travel 0, 2062, 2133, 2908 miles with respective probabilities 0.5, 0.5, 0, 0. So the expected length of a one-day trip from SF is

$$0 \times 0.5 + 2062 \times 0.5 + 2133 \times 0 + 2908 \times 0 = 1031.$$

Similarly, we find the expected length of a one-day trip from STL and CHI to be 709.8 and 804.3 miles, respectively.

Last, using a standard result for Markov Chains, we can compute the expected length of the trip from each starting city by multiplying our matrix E by the column of expected length of one-day trips.

$$\begin{bmatrix} 10 & 10 & 4 \\ 8 & 10 & 4 \\ 6 & 6.67 & 4 \end{bmatrix} \times \begin{bmatrix} 1031 \\ 709.8 \\ 804.3 \end{bmatrix} = \begin{bmatrix} 20{,}625 \\ 18{,}563 \\ 14{,}135 \end{bmatrix}$$

The product vector gives the expected length in miles of the total trip starting from each of the three cities. So if we start at San Francisco, we can expect to log 20,625 frequent flyer miles before getting to New York City. We will travel, on average, 18,563 miles if we start our trip at St. Louis.

Case Study 9-2: A Half-Inning of Baseball as a Markov Chain

Topics Covered: State of an inning in a baseball game, transitions between states, batting plays, runs scored   In the first case study, we described the basic structure of a Markov Chain. How can this model be applied to baseball?

We focus on the run production of a baseball team during its half-inning at-bat. A state will describe the current runners on bases situation and the number of outs. Each base (first, second, and third) can be either occupied or not. A total of eight possible base situations are graphically represented in Table 9.3 using symbols on a diamond shape.

TABLE 9.3
Possible runner situations in a baseball game.

Bases situation	Empty	Runner on first	Runner on second	Runner on third	Runners on 1st and 2nd	Runners on 1st and 3rd	Runners on 2nd and 3rd	Bases loaded
Symbol	◇	◇	◇	◇	◇	◇	◇	◇

When a batter comes up, the number of outs at any one time during an inning can either be 0, 1, or 2. So when a batter comes to bat, there are eight possible base situations and three different out situations, and so there are $8 \times 3 = 24$ different base/out situations. If we add the final situation, three outs, to this list, we have a total of $24 + 1 = 25$ possible states. In Table 9.4 all of the states are presented by use of a table classified by number of outs and the base situation.

A player comes to bat when the inning is in a particular state, say runners on 1st and 2nd with one out. There will be a *batting play*, such as a hit, out, or walk, that will change

TABLE 9.4

Diagram of all 25 possible states of an inning defined by the runners on base situation and the number of outs.

Bases Situation							
◇	◇	◇	◇	◇	◇	◇	◇

0 outs

1 out

2 outs

3 outs

the inning state. For example, suppose a batter comes up with a runner on 1st and one out

$$\left(\diamondsuit, 1 \text{ out} \right)$$

He singles into right field, moving the runner from 1st to 3rd. The new inning state is

$$\left(\diamondsuit, 1 \text{ out} \right)$$

We can view a half-inning of baseball as a sequence of changes in state, starting with (no runners on base, 0 out) and ending with (3 outs). To simplify the following discussion, we will ignore *non-batting plays*, such as steals and balks,[2] that can also change the state of an inning. A more realistic model would include these non-batting plays in the analysis.

A Markov Chain can be used to model the change in states during an inning. We assume that the probability of moving to a particular (bases, outs) state depends only on the current state and not on any earlier state. We let p_{ij} denote the probability of moving from the ith state to the jth state. We let P denote the matrix of transition probabilities p_{ij} with 25 rows and 25 columns. Note that the (3 outs) situation is an absorbing state in the Markov Chain, since one cannot leave this state once it is entered. In other words, the inning is over when there are three outs.

When there is a change in the (bases, outs) state, runs can score. Suppose that there are R_{before} runners on base and O_{before} outs before the batting play, and R_{after} runners and O_{after} outs after the play. Then the number of runs scored in this transition would be

$$\text{Runs Scored} = (R_{\text{before}} + O_{\text{before}} + 1) - (R_{\text{after}} + O_{\text{after}}).$$

Let's illustrate this computation for one batting play. Suppose there are runners on 1st and 2nd with one out. The batter doubles, scoring both runners, leaving a runner on 2nd with one out. Table 9.5 verifies that the Runs Scored formula gives that two runs scored on this play.

---

[2]A balk is when the pitcher stops during his pitching motion and all runners advance one base.

TABLE 9.5
Illustration of the number of runs scored for one batting play.

Situation before play	(R_{before}, O_{before})	Play	Situation after play	(R_{after}, O_{after})	Runs scored on play
◇ , 1 out	(2, 1)	Double	◇ , 1 out	(1, 1)	$(2 + 1 + 1)$ $- (1 + 1) = 2$

Case Study 9-3: Useful Markov Chain Calculations

Topics Covered: Computation of probabilities by use of frequencies, matrix computations, expected number of visits to states, expected runs scored in a half-inning, computation of event probabilities by simulation   To use this Markov Chain model, we need to estimate the matrix of transition probabilities P. We use play-by-play data for all the National League teams in the 1987 season to estimate this matrix.

Let's illustrate computing the transition probabilities starting from the bases empty, no outs state. All half-innings begin with this particular state, although this state may happen more than once in a particular inning. In the 1987 National League season, there were 18,099 instances where this state occurred. There are five possible transitions from this state, depending on the batting play:

• The batter can hit a home run, and the state remains at bases empty and no outs.

• The batter can get out, and the state changes to bases empty and one out.

• The batter can get to first base by a single, walk, hit-by-pitch, or an error and the state changes to runner on first and no outs.

• The batter can hit a double, and the state changes to runner on second with no outs.

• The batter can hit a triple, and the state changes to runner on third base with no outs.

Table 9.6 below shows the five possible transitions from (bases empty, no outs) and the frequency with which each transition occurred. Note that the most common transition is an out—this event happened 12,052 times out of a total of 18,099 transitions and so the probability of this transition is estimated to be 12052/18099 = .6659. In a similar fashion, we can estimate the probability of the four other possible transitions.

Suppose instead that the current state is two outs with a runner on first. In this case, there are more possible transition types depending on the advancement of the runner from first base. Table 9.7 shows all of the possible transitions from (runner on first, two outs), the frequency of each transition, and the corresponding probability. We see that the most likely play is an out that results in the third out—the probability of this transition is 2944/4380 = .6721. Other possible transitions, ordered in terms of their likelihood of occurring, are

• a single, walk, or hit-by-pitch that result in runners on 1st and 2nd,

• a single where the runner on first advances to 3rd base,

• a home run that clears the bases,

TABLE 9.6

All possible transitions from the no-runners, no outs state. For each state, the number of runs scored, the number of times this transition occurred, and the corresponding probability are given.

Beginning state	Batting play	End state	Runs scored	Count	Probability
, 0 out	Home run	, 0 out	1	489	.0270
	Out	, 1 out	0	12052	.6659
	Single or walk or hit-by-pitch or error	, 0 out	0	4584	.2533
	Double	, 0 out	0	845	.0467
	Triple	, 0 out	0	129	.0071
Total				18099	1.0000

TABLE 9.7

All possible transitions from the (runner on first, two outs) state. For each state, the number of runs scored, the number of times this transition occurred, and the corresponding probability are given.

Beginning state	Batting play	End state	Runs scored	Count	Probability
, 2 outs	Home run	, 2 outs	2	127	.0290
	Single	, 2 outs	1	1	.0002
	Double	, 2 outs	1	95	.0217
	Triple	, 2 outs	1	35	.0080
	Single or walk or hit-by-pitch	, 2 outs	0	830	.1895
	Single	, 2 outs	0	253	.0578
	Double	, 2 outs	0	95	.0217
	Out	3 outs	0	2944	.6721
Total				4380	1.0000

- a double that scores the baserunner on 1st,
- a double that advances the baserunner to 3rd base,
- a triple that scores the baserunner on 1st,
- a single that scores the baserunner on 1st.

We see that the single scoring the runner on first is quite an unusual play—it happened only one time that particular season.

   Suppose that we compute the probabilities of all possible transitions starting from each of the 24 possible initial states. We can then construct the matrix P (of dimension 25 by 25) that contains the transition probabilities for all (runners on base, number of outs) states. Given this transition matrix, we can perform several matrix calculations, such as done in the first case study, to find some quantities of interest.

Reaching Various States after a Given Number of Batters

Suppose three players come to bat at the beginning of an inning. What will be the state of the inning after these three plate appearances? What's the chance that there will be at least one runner in scoring position?

   Remember the matrix P gives the probabilities of one-step transitions. To obtain the probabilities of three-step transitions, we simply multiply the matrix by itself three times:

$$P^3 = P \times P \times P.$$

Table 9.8 gives the first row of the matrix P^3, displaying it in a familiar two-way format where rows correspond to the number of outs and the columns represent the runners situation.

TABLE 9.8
Transition probabilities of being in different states after three plays starting from the (no outs, no runners) state.

	\diamond	\diamond (1st)	\diamond (2nd)	\diamond (3rd)	\diamond (1st,2nd)	\diamond (1st,3rd)	\diamond (2nd,3rd)	\diamond (loaded)
0 outs	.0025	.0035	.0027	.0008	.0077	.0052	.0023	.0098
1 out	.0150	.0260	.0181	.0075	.0850	.0335	.0209	0
2 outs	.0446	.2367	.0859	.0321	0	0	0	0
3 outs				.3601				

   These probabilities represent the chances of being in different inning states after three batters starting from the (no outs, no runners) state. We see that the most likely state after three at-bats is (3 outs) with a probability of .3601. The next most likely state is (2 outs, runner on first) with a probability of .2367. Several three-batter movements are impossible. For example, we see from the table that the probability of getting to (2 outs, runners on first and second) from three batters has a probability of 0.

What is the chance of having a runner in scoring position (that is, a runner on 2nd or 3rd base) after three at-bats? We look at the above table and sum the probabilities over the

◇, ◇, ◇, ◇, ◇, ◇

states where one runner or more are in scoring position. So Prob(runner in scoring position after 3 batters) $= .0027 + .0181 + .0859 + \cdots + 0 = .3116$.

How Many Batters?

As before, let Q denote the submatrix of P found by deleting the one row and one column corresponding to the absorbing "3 outs" state. The matrix $E = (I - Q)^{-1}$ contains the expected times that the inning will be in each state starting with each of the 24 possible beginning states. Suppose that we are currently at the (no runners on, no outs) state that begins the inning. The first row of the matrix E will give the expected number of visits to all the states (before absorption) given that one starts in this (no runners on, no outs) state. This first row of E is displayed in a convenient table format in Table 9.9.

TABLE 9.9

The expected number of visits to each state starting from the (no runners on, no outs) state.

	Bases Situation							
	◇	◇	◇	◇	◇	◇	◇	◇
0 outs	1.039	0.269	0.058	0.011	0.063	0.027	0.014	0.014
1 out	0.728	0.305	0.118	0.034	0.118	0.054	0.034	0.042
2 outs	0.570	0.312	0.137	0.056	0.152	0.075	0.039	0.051
Sum								4.317

Of course, since we are starting at the (no runners on, no outs) state, we'll visit this state at least once—this table tells us that we'll visit it, on the average, 1.039 times. Also, we will visit the (runner on first, no outs) state an average of .269 times, the (runner on 2nd, no outs) state an average of .058 times, and so on.

If we sum all of these expected state counts, we will obtain the expected number of state visits before absorption. In baseball lingo, this sum will be the expected number of batters before the inning is over. Here the sum is 4.317, which means that, on average, there will be 4.317 batters in the remainder of the inning starting with the (no runners on, no outs) state.

In a similar fashion, one can use the matrix E to compute the expected number of batters in the remainder of the inning starting from each of the 24 possible states. Table 9.10 shows the "expected number of batters" matrix.

TABLE 9.10
Expected number of batters in the remainder of the inning starting from each possible state.

	Bases Situation							
0 outs	4.317	4.079	4.389	4.332	4.011	4.045	4.422	3.919
1 out	2.905	2.719	2.977	2.993	2.664	2.732	3.116	2.652
2 outs	1.491	1.484	1.567	1.569	1.434	1.478	1.635	1.406

A manager can use this matrix in strategic decisions during a game. For example, suppose there are bases loaded with one out. Looking at this table, we see that, on average, there will be 2.652 batters in the remainder of this inning. A manager can use this to plan his batting lineup; for example, it might help him make a decision regarding the use of a pinch-hitter.

Expected Runs in the Remainder of the Inning

The expected number of visits matrix E can be used to compute the expected number of runs in the remainder of the inning starting from each state. This computation is similar to the computation for the expected number of frequent flyer miles for our first example. Let $R_{\text{one step}}$ denote the column vector that contains the expected number of runs that will be scored in a *single* batting play starting from each of the 24 states. Let's illustrate the computation of the first element of $R_{\text{one step}}$. Suppose that one starts in the (no runners on, no outs) state. In one batting play, only 0 and 1 runs can score. The probability of scoring one run is the chance of hitting a home run that is estimated to be .027, and so the chance of scoring no runs is estimated to be $1 - .027 = .973$. The expected number of runs scored in one batting play is therefore

$$1 \times .027 + 0 \times .973 = .027.$$

If we do this computation for each of the 24 states, we obtain the vector $R_{\text{one step}}$, displayed in matrix form in Table 9.11.

The vector of expected number of runs scored, denoted by R, is found by multiplying the number of visits matrix E by $R_{\text{one step}}$,

$$R = E \times R_{\text{one step}}.$$

TABLE 9.11

The expected number of runs scored in a single batting play starting at each possible state. The vector $R_{\text{one step}}$ contains these values.

	Bases Situation							
0 outs	.027	.070	.148	.496	.238	.602	.629	.887
1 out	.024	.073	.163	.502	.246	.572	.492	.798
2 outs	.024	.088	.177	.258	.256	.337	.306	.516

Note that R is a column vector with 24 entries, where the entries correspond to the expected number of runs in the remainder of the inning starting from each of the 24 states. This vector R is displayed, in matrix form, in Table 9.12.

TABLE 9.12

The expected number of runs scored in the remainder of the inning starting at each possible state.

	Bases Situation							
0 outs	0.48	0.85	1.08	1.35	1.38	1.63	1.90	2.14
1 out	0.26	0.50	0.65	0.94	0.85	1.13	1.32	1.50
2 outs	0.10	0.22	0.31	0.38	0.41	0.49	0.54	0.66

This is a fundamental matrix that is useful for many purposes in baseball research. Starting an inning in the (no runners, no outs) state, we see from Table 9.12 that a team will score, on average, .48 runs. In contrast, when bases are loaded with one out, there is a good potential to score runs; using the table, we see that, on average, 1.50 runs will score from this state. We will refer to this matrix as the *run potential matrix*, since it gives the potential for a team scoring runs starting from each (bases, outs) situation.

Computing Other Event Probabilities

Once the Markov Chain model is defined by means of the transition probability matrix P, one can compute the probability of any event of interest by simulating the chain many times. For example, suppose a manager is interested in the probability that the team will score at least one run if there is a runner on 3rd base with one out. On a computer, one can simulate the batting plays in the remainder of the inning by using the transition probability matrix, starting with the row of the matrix corresponding to the (3rd base, one out) state, and stopping when the (3 outs) state has been reached. We record if one or more runs were scored in this simulated inning. Then we repeat this simulation process for a large number of innings, each time recording if one or more runs occurred. Then the probability of scoring at least one run, Prob(at least one run), can be approximated by

$$\frac{\text{number of innings where at least one run was scored}}{\text{number of innings simulated}}.$$

We actually simulated the Markov Chain starting from each of the 24 possible inning states. For each starting state, we simulated the remainder of the inning 10,000 times. From each state, we computed the probability that at least one run is scored in the remainder of the innings. The estimated probabilities are displayed in Table 9.13.

TABLE 9.13
The probability of scoring at least one run in the remainder of the inning starting from each possible state.

	Bases Situation							
0 outs	.265	.405	.598	.842	.592	.814	.813	.824
1 out	.149	.250	.382	.665	.394	.608	.633	.641
2 outs	.070	.126	.207	.277	.215	.256	.226	.275

When the inning starts (no outs, no runners on), the probability the team will score at least one run is .265. This chance of scoring drops to .070 when there are two outs in the inning. A team is most likely to score when the bases are loaded with no outs—the chance of scoring is .824.

Case Study 9-4: The Value of Different On-base Events

Topics Covered: Value of a batting event defined by expected runs, mean value of a particular type of hit   Using the run potential matrix R defined in the previous case study, we can define the value of a batting event. Suppose a batter comes to bat in a particular state (runners on base and number of outs) where the expected number of runs scored in the remainder of the inning is R_{before}. After the batter event, there is a new (runners on base, outs) situation with an expected number of runs scored called R_{after}. Then the value of this batting event is

$$\text{Value} = R_{after} - R_{before} + (\text{runs scored on play}).$$

Values of a Terrible Play and a Great Play

Let us illustrate this formula for two extreme cases that correspond to the least and most valuable batting plays. Suppose that the bases are loaded with no outs. The batter hits a sharp grounder to third; the third baseman touches third base, throws it quickly to the second baseman who touches second base and throws it to first, completing an (unusual) triple play. Clearly this is a bad play for the hitter (and the team)—the question is how bad?

When the batter came to the plate, the run potential (bases loaded, no outs) is 2.14 runs. After the play, there are three outs and a run potential of zero—no runs scored on this play. The value of this plate appearance is

$$\text{Value} = (0 - 2.14) + 0 = -2.14.$$

So this bad hitting play essentially cost the team about two runs. This is the worst possible play where the worth is defined in terms of this value measure.

Let's contrast this with a great batting play. The bases are loaded with two outs and the batter hits a deep fly that goes over the center field fence—it's a grand slam! When this batter came to bat, the run potential of (bases loaded, two outs) is .66 runs. After the play, the bases are empty (with 2 outs) and the run potential is .10 runs. Four runs scored on this play. The value of this plate appearance is

$$\text{Value} = (.10 - .66) + 4 = 3.44.$$

One might think the value of this home run would be four runs—after all, four runs scored on this play. But, by clearing the bases, the batter has decreased the run potential in the remainder of the inning, and the value measure adjusts for this decrease.

The Value of a Home Run

Using the notion of value, we can measure the effectiveness of different types of batting plays, such as hits, walks, sacrifice flies, and outs. Here we focus on the biggest hit, the home run. There were a total of 1797 home runs hit in the National League in 1987. But these home runs were not equally valuable—certainly a home run hit with runners on base is more valuable than a home run hit with the bases empty. In Table 9.14, we classify all of the home runs by the bases situation and the number of outs. In each cell, the number on

TABLE 9.14
Values and the number of home runs that occur in all possible situations.

	Bases Situation							
	◇	◇	◇	◇	◇	◇	◇	◇
0 outs	1	1.64	1.41	1.14	2.11	1.85	1.58	2.35
	(489)	(93)	(25)	(4	(23)	(9)	(7)	(8)
1 out	1	1.76	1.60	1.31	2.40	2.12	1.94	2.76
	(308)	(103)	(54)	(16)	(37)	(27)	(7)	(16)
2 outs	1	1.88	1.79	1.72	2.69	2.61	2.56	3.44
	(253)	(127)	(51)	(30)	(47)	(33)	(11)	(19)

top is the value of the home run in that particular situation and the number in parentheses is the number of home runs hit in that situation.

Note that most of the home runs were hit with the bases empty—in fact, 1050 (58%) of the home runs hit were solo shots and the value of each of these home runs is 1. In contrast, only 14% of the home runs were hit with two or more runners on base. The values of these "two runners or more on base" home runs vary from 1.58 (runners on 2nd and 3rd and no outs) to 3.44 (bases loaded with 2 outs).

To measure the value of a home run, we average all of the values for the 1797 home runs hit in the 1987 National League season. To find this average, we first total all of the values of the home runs in Table 9.14, and divide this total by the number of home runs. In Table 9.15, we find the total value in each situation, and show the sum of these totals in the lower right cell of the table. The average value of a home run is then equal to

$$\text{Average value of home run} = \frac{489 \times 1 + 93 \times 1.64 + 25 \times 1.41 + \cdots + 19 \times 3.44}{1797}$$

$$= \frac{2517.50}{1797} = 1.40.$$

TABLE 9.15
Illustration of the computation of the average value of a home run.

	Bases Situation							
	(empty)	(1st)	(2nd)	(3rd)	(1st,2nd)	(1st,3rd)	(2nd,3rd)	(loaded)
0 outs	489×1 $= 489$	93×1.64 $= 152.52$	25×1.41 $= 35.25$	4×1.14 $= 4.56$	23×2.11 $= 48.53$	9×1.85 $= 16.65$	7×1.58 $= 11.06$	8×2.35 $= 18.80$
1 out	308×1 $= 308$	103×1.76 $= 181.28$	54×1.60 $= 86.40$	16×1.31 $= 20.96$	37×2.40 $= 88.80$	27×2.12 $= 57.24$	7×1.94 $= 13.58$	16×2.76 $= 44.16$
2 outs	253×1 $= 253$	127×1.88 $= 238.76$	51×1.79 $= 91.29$	30×1.72 $= 51.60$	47×2.69 $= 126.43$	33×2.61 $= 86.13$	11×2.56 $= 28.16$	19×3.44 $= 65.36$
								Total $= 2517.50$

This average value of a home run may seem small, since we credit a home run with four bases in the computation of a slugging percentage. But this average of 1.40 runs represents a typical run value for this type of hit. In the exercises, we will investigate the run values for other types of batting plays.

Case Study 9-5: Answering Questions About Baseball Strategy

Topics Covered: Evaluating the worth of a play by use of expected runs   The 2001 World Series between the New York Yankees and the Arizona Diamondbacks is considered one of the most exciting World Series in history. We focus on Game 4, which involved repeated use of a well-known baseball strategy. The question we want to address is whether this strategy really helps in scoring runs.

In the top of the first inning, the lead-off hitter for Arizona, Tony Womack, singled to center. Then Craig Counsell, the second batter, was instructed to hit a sacrifice bunt. The bunt was effective—Counsell was thrown out at first and Womack advanced to second base. The inning finished without Womack scoring. In the top of the third inning, the same situation happened. Womack opened the inning with a walk and Counsell again sacrificed with a bunt to move Womack to second. (Again Womack didn't score.) In the top of the 5th inning, Womack started the inning with a double. Counsell again hit a sacrifice bunt, moving Womack to third. The next hitter, Luis Gonzalez, hit a fly ball, but Womack was thrown out at home plate, ending the inning.

It appears that the Arizona manager, Bob Brenly, likes to play the sacrifice bunt. Counsell was instructed to sacrifice his at-bat (and get an out) in order to advance the runner one additional base. Is the sacrifice bunt a smart play in baseball?

We have the tools to evaluate the effectiveness of this play in creating runs for a team. As in evaluating the home run, we use the run potential matrix that gives the average number of runs scored in each bases/outs situation.

Sacrifice Bunt to Move Runner from First to Second

We first look at the situation that occurred in the first and third innings of the World Series game. Arizona has a runner on first with no outs. Looking at our run potential matrix, this situation has a potential of .85 runs. If Counsell hits a successful sacrifice bunt, the batter is out, but the runner moves to second base. The run potential of (runner on 2nd, one out) is .65 runs. So the value of the sacrifice bunt in this situation is

$$\text{Value} = 0.65 - 0.85 = -0.20$$

So Arizona really has hurt themselves—on the average the team has decreased their run production by .2 runs.

But this calculation is assuming that we are primarily interested in scoring as many runs as possible. Maybe the Arizona manager wants to score one run or more and thinks that he has put the team in a better position to score (that is, get one run or more) using the sacrifice bunt. Using the Markov Chain model, we showed earlier how we could simulate the probability of scoring in the remainder of the inning from each of the 24 starting states. From Table 9.13, we obtain

$$\text{Prob(scoring with runner on 1st and no outs)} = .405,$$

$$\text{Prob(scoring with runner on 2nd and 1 out)} = .382.$$

Even from this perspective, Arizona has decreased their chances of scoring slightly by using the sacrifice bunt.

Let's return to the situation where these sacrifice bunts occurred. Counsell was instructed to sacrifice in the 1st and 3rd innings of the game. In these early innings, it would seem that the goal of a team would be to score as many runs as possible. Also, Craig

Counsell is one of the better hitters on the Diamondbacks and he is likely to get a base hit that would greatly increase Arizona's run potential. So it would seem that the sacrifice hit wasn't an effective strategy in the situations where it was used.

Exercises

Leadoff Exercise. Rickey Henderson is considered the greatest leadoff hitter, but how often did he actually lead off? Let's focus on Rickey's 273 plate appearances at home games for the 1990 baseball season—we record the state of the inning (outs and runners) for each plate appearance. Table 9.16 classifies these 273 plate appearances with respect to the number of outs (0, 1 and 2) and the eight possible runner situations.

TABLE 9.16
Count of number of plate appearances in different runner and out situations for home games for the 1990 season.

	Bases Situation							
0 outs	109	17	0	0	2	2	0	0
1 out	29	13	8	1	3	2	0	2
2 outs	37	18	11	4	6	5	2	2

TOTAL = 273

(a) What fraction of times did Rickey actually bat with the bases empty and no outs? (These were essentially the times when he was a leadoff batter.)

(b) What fraction of times did Rickey bat with exactly one runner on base? With exactly two runners on base? When the bases were loaded?

(c) Suppose that you were able to find a similar classification of plate appearances for Barry Bonds for the 1990 season. Would you expect Bonds to have similar fractions of plate appearances with the bases empty, one runner, two runners, and three runners as Rickey Henderson? Explain.

9.1. A baseball fan has been celebrating the recent success of his team at location D. He takes a random walk down the street hoping to arrive at his home (location H). From location D, he is sure to go to location J in the next minute. From location J, he is equally likely in the next minute to return to D or go ahead to location K. From location K, he is equally likely the next minute to go home (location H) or back to location J. Once he is home, he will remain there with probability 1. Figure 9.3 shows the four locations, arrows to show the possible transitions, and the transition probabilities. A Markov Chain with states D, J, K, H and transition matrix P shown in Table 9.17 can represent this random walk. (Note that H is an absorbing state in the chain.) The matrices P^2, P^3, and P^4 are also shown below.

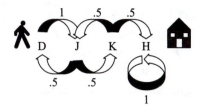

FIGURE 9.3
States and possible moves in a random walk from location D to home (H).

TABLE 9.17
Transition probability matrix for the random walk.

$$
P = \begin{array}{c|cccc}
 & D & J & K & H \\
\hline
D & 0 & 1 & 0 & 0 \\
J & 0.5 & 0 & 0.5 & 0 \\
K & 0 & 0.5 & 0 & 0.5 \\
H & 0 & 0 & 0 & 1
\end{array}
$$

$$
P^2 = \begin{bmatrix} .5 & 0 & .5 & 0 \\ 0 & .75 & 0 & .25 \\ .25 & 0 & .25 & .50 \\ 0 & 0 & 0 & 1 \end{bmatrix}
\quad
P^3 = \begin{bmatrix} 0 & .75 & 0 & .25 \\ .375 & 0 & .375 & .25 \\ 0 & .375 & 0 & .625 \\ 0 & 0 & 0 & 1 \end{bmatrix}
$$

$$
P^4 = \begin{bmatrix} .375 & 0 & .375 & .25 \\ 0 & .5625 & 0 & .4375 \\ .1875 & 0 & .1875 & .6250 \\ 0 & 0 & 0 & 1 \end{bmatrix}
$$

(a) One possible path of our baseball fan is DJDJKH. Find the probability of this path using the transition probabilities.

(b) If the fan starts at D, is it possible for our fan to return to D after three minutes? (Recall that each step takes one minute.) Why or why not?

(c) Using the given matrices, find the probability that the fan (starting at D) will arrive home in four minutes.

(d) If the fan starts at J, find the probability that he returns to J in two minutes.

(e) If the fan starts at J, where is his most likely location in three minutes?

9.2. (Exercise 9.1 continued) Let Q denote the matrix obtained by deleting the last row and last column corresponding to the absorbing state from the transition matrix P. The fundamental matrix $E = (I - Q)^{-1}$ is displayed below.

$$
E = (I - Q)^{-1} = \begin{bmatrix} 3 & 4 & 2 \\ 2 & 4 & 2 \\ 1 & 2 & 2 \end{bmatrix}.
$$

   (a) If the fan starts at D, how many minutes does the typical fan expect to spend at location J before he arrives home?

   (b) If the fan starts at J, how many minutes will the typical fan spend, on average, at location D before he gets home?

   (c) Starting from D, how many minutes will the fan take, on average, before arriving home?

   (d) Is it possible for the fan to take 20 minutes to arrive home (starting from D)? Why or why not?

9.3. Consider the following simplified version of baseball. When a player comes to bat, he gets out with probability .6, he hits a double with probability .3, and he hits a home run with probability .1. No other batting events besides outs, doubles, and home runs are possible. There are only seven possible (outs, bases) situations in this game, given by

$$\left(0 \text{ outs}, \diamondsuit\right), \left(0 \text{ outs}, \blacklozenge\right), \left(1 \text{ out}, \diamondsuit\right), \left(1 \text{ out}, \blacklozenge\right),$$

$$\left(2 \text{ outs}, \diamondsuit\right), \left(2 \text{ outs}, \blacklozenge\right), (3 \text{ outs})$$

   (a) Fill in the transition probability matrix P below. The impossible transitions are indicated by zeros in the matrix.

Transition matrix P

	0 outs, ◇	0 outs, ◆	1 out, ◇	1 out, ◆	2 outs, ◇	2 outs, ◆	3 outs
0 outs, ◇				0	0	0	0
0 outs, ◆				0	0	0	0
1 out, ◇	0	0				0	0
1 out, ◆	0	0			0		0
2 outs, ◇	0	0	0	0			
2 outs, ◆	0	0	0	0			
3 outs	0	0	0	0	0	0	

   (b) For each of the possible transitions, find the number of runs scored on the play and place in the table below. The impossible transitions are crossed out in the table.

Runs matrix

	0 outs, ◇	0 outs, ◆	1 out, ◇	1 out, ◆	2 outs, ◇	2 outs, ◆	3 outs
0 outs, ◇				xxx	xxx	xxx	xxx
0 outs, ◆				xxx	xxx	xxx	xxx
1 out, ◇	xxx	xxx				xxx	xxx
1 out, ◆	xxx	xxx				xxx	xxx
2 outs, ◇	xxx	xxx	xxx	xxx			
2 outs, ◆	xxx	xxx	xxx	xxx			
3 outs	xxx	xxx	xxx	xxx	xxx	xxx	

9.4. (Exercise 9.3 continued)   Consider a Markov Chain model for the simplified game of baseball with transition matrix found in Exercise 9.3.

(a) Find the two-step probability matrix P^2. Using this matrix, find the probability that there will be a runner on 2nd and one out after two players have batted in the inning.

(b) Compute the expected number of visits matrix E. Using this matrix, find the average number of batters in a half-inning of baseball.

(c) If the current state of the inning is one out with a runner on 2nd, find the expected number of batters in the remainder of the inning.

(d) Find the run potential vector $R = E \times R_{\text{one step}}$. This vector gives the expected number of runs in the remainder of the inning starting at each possible state.

(e) Suppose that a player comes to bat with a runner on 2nd with one out. He hits a double. Using the run potential vector R, find the value of this play.

9.5. Consider again the run potential matrix shown in Table 9.18 that gives the expected runs in the remainder of the inning for each of the 24 possible outs/bases situations.

TABLE 9.18
The expected number of runs scored in the remainder of the inning starting at each possible state.

	Bases Situation							
	◇	◇	◇	◇	◇	◇	◇	◇
0 outs	0.48	0.85	1.08	1.35	1.38	1.63	1.90	2.14
1 out	0.26	0.50	0.65	0.94	0.85	1.13	1.32	1.50
2 outs	0.10	0.22	0.31	0.38	0.41	0.49	0.54	0.66

Using this matrix, find the value of the following batting plays.

(a) There are runners on the corners (first and third) with one out. The batter hits a double, scoring both runners.

(b) There is a runner on 1st with no outs. The batter hits a grounder, which is converted to a double play, getting both the runner and the batter out.

(c) The bases are loaded with no outs. The batter hits a grand slam home run. Compare with the value of the grand slam with two outs.

9.6. What is the value of a single when there are runners on 1st and 2nd with no outs? The value of this hit depends on the advancement of the runners. In the 1987 National League, a single occurred in this situation (runners on 1st and 2nd with no outs) a total of 136 times. Table 9.19 shows the three types of run advancement and the count of each type.

TABLE 9.19
Three types of runner advancement and the count of each type when there is a single with runners on 1st and 2nd and no outs.

Starting state	Final state	Count	Runs scored	Value
	, no outs	43		
, no outs	, no outs,	35		
	, no outs	58		

(a) For each final state, compute the number of runs scored on the play. Put these values in the "Runs scored" column of the table.

(b) Using the run potential matrix in Table 9.12 and the runs scored from (a), find the value of each transition. Put the values in the "Value" column of the table.

(c) Use the previous calculations to find the mean value of a single when runners are on 1st and 2nd with no outs.

9.7. Suppose that the batter hits a single with the bases loaded and one out. How important is this play? The value depends on the advancement of the runners. This play occurred 88 times in the 1987 National League. Table 9.20 below shows the possible advancement of the runners and the number of times each type of advancement occurred.

(a) For each final state, compute the number of runs scored on the play. Put these values in the "Runs scored" column of the table.

(b) Using the run potential matrix and the runs scored from (a), find the value of each transition. Put the values in the "Value" column of the table.

TABLE 9.20
Five types of runner advancement and the count of each type when there is a single with the bases loaded with one out.

Starting state	Final state	Count	Runs scored	Value
(bases loaded), 1 out	(diagram), 1 out	2		
	(diagram), 1 out	34		
	(diagram), 1 out	14		
	(diagram), 1 out	5		
	(diagram), 1 out	33		

(c) Use the previous calculations to find the mean value of a single when the bases are loaded with one out.

9.8. How valuable is a walk? The value of this play depends on the beginning (runners, outs) state. In Table 9.21, the value of the walk is shown for each of the 24 possible states and the number of times that the state occurred is displayed in parentheses. For example, we see that there was a walk with the bases empty and no outs 1468 times. The value of this particular play, moving from

(bases empty), 0 outs to (runner on first), 0 outs

is .3624 runs.

TABLE 9.21
Values and the number of walks that occur in all possible situations.

	Bases Situation							
	◇	◇	◇	◇	◇	◇	◇	◇
0 outs	0.36 (1468)	0.53 (248)	0.30 (130)	0.29 (31)	0.76 (54)	0.50 (28)	0.23 (35)	1.00 (8)
1 out	0.24 (1035)	0.35 (290)	0.20 (387)	0.19 (112)	0.65 (128)	0.36 (66)	0.18 (196)	1.00 (46)
2 outs	0.12 (983)	0.19 (347)	0.10 (606)	0.11 (192)	0.25 (173)	0.17 (83)	0.12 (219)	1.00 (57)

(a) In what situation(s) is a walk most valuable? When is a walk least valuable?

(b) Explain why a walk has a value of one run when the beginning state is "bases loaded."

(c) In what situation is a walk most likely to occur? When is a walk least likely to occur?

(d) Find the average value of a walk over all situations.

9.9. In Game 7 of the 2001 World Series, Steve Finley and Mark Grace were starters for the Arizona Diamondbacks. The tables below show how the players did in all of their plate appearances in that particular game. Table 9.22 gives the inning in which the player batted, the before and after game states and the batting play.

TABLE 9.22

Results of all plate appearances of Steve Finley and Mark Grace in Game 7 of the 2001 World Series.

Steve Finley

Inning	Before state	Play	After state	Value
2nd	, no outs	Ground out to shortstop	, 1 out	
3rd	, 2 outs	Struck out	3 outs	
6th	, no outs	Single to center	, no outs	
8th	, 2 outs	Single to right	, 2 outs	

Mark Grace

Inning	Before state	Play	After state	Value
2nd	, 1 out	Single to left	, 1 out	
4th	, 1 out	Single to left	, 1 out	
6th	, 1 out	Ground out to second	, 2 out	
9th	, no outs	Single to center	, no outs	

(a) For each batting play of each player, find the value using the run potential matrix in Table 9.12. Record the values in the Value columns of the tables.

(b) What was the most valuable batting play for each hitter?

(c) Which player had the better batting performance in this particular game? Explain.

9.10. (Value of a sacrifice bunt)   Suppose that there is a runner on 2nd base with no outs. Is it a good play to have the batter hit a sacrifice bunt to move the runner from 2nd to 3rd? Use the run potential matrix in Table 9.12 in your explanation. Alternately, you can use the matrix that gives the probability of scoring at least one run in all situations.

9.11. (Value of a steal)   Suppose the first batter in the inning gets a walk. He is thinking about stealing 2nd base.

(a) Suppose that the runner attempts to steal 2nd base and is successful. Find the value of this play. (It should be positive.)

(b) Suppose that the runner attempts to steal 2nd base and is unsuccessful (the catcher throws him out). Find the value of this play. (This value should be negative.)

(c) Suppose that this particular runner is successful in stealing 2nd base 70% of the time. Does the team benefit (in terms of runs scored) by having this player steal? Explain.

(d) Let the probability that the runner is successful in stealing 2nd base be equal to p. Find the value of p such that the value of the stealing play is equal to zero. (So if the runner has a success rate that is larger than p, it would benefit the team to have him attempt the steal of 2nd base.)

Further Reading

A good description of the discrete Markov Chain probability model is contained in Kemeny and Snell (1976). Pankin (1987) and Bukiet and Palacios (1997) describe the use of Markov Chains to model baseball. Lindsey (1963), using actual game data, found the distribution of runs scored in the remainder of the inning starting from each possible bases occupied/outs situation. Lindsey's run production data is used to estimate the value of different types of base hits in Albert and Bennett (2001), Chapter 7.

A

An Introduction to Baseball

Baseball is one of the most popular games in the United States; it is often called the national pastime. The game evolved out of various ball-and-stick games played in many areas of the world, including the Russian game of lapta and the English game of rounders. It became a popular sport in the eastern United States in the mid-1800s. Professional baseball started near the end of the 19th century; the National League was founded in 1876 and the American League in 1900. Currently in the United States, there are 30 professional teams in the American and National Leagues and millions of people watch games in ballparks and on television.

Baseball is a game between two teams of nine players each, played on an enclosed field. A game consists of nine innings. Each inning is divided into two halves; in the top half of the inning, one team plays defense in the field and the second team plays offense, and in the bottom half, the teams reverse roles. The fielding positions for the nine players playing defense are catcher, first baseman, second baseman, shortstop, third baseman, left fielder, center fielder and right fielder. Figure A.1 shows a diagram of a baseball field and shows the fielding positions for the defensive players. In addition, this figure shows the four bases—home base, first base, second base, and third base—that play an important part in the game. This figure also shows the location of one offensive player, the batter, at home base.

The team batting during a particular half-inning, the offensive team, is trying to score *runs*. A player from the offensive team begins by batting at home base. A run is the score made by this player who advances from batter to runner and touches first, second, third, and home bases in that order. A team wins a game by scoring more runs than its opponent at the end of nine innings. There are some exceptions to the nine-inning game. A game that is tied after nine innings continues into extra innings until one team has won, and a game may be shortened due to inclement weather.

A basic play in baseball consists of a player on the defensive team, the pitcher, throwing a spherical ball (called a pitch) toward the batter. The batter attempts to strike or hit the

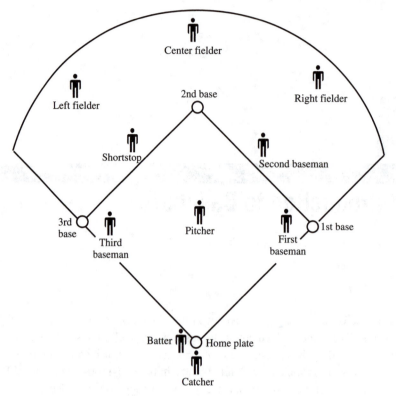

FIGURE A.1

Diagram of baseball field and bases and location of nine defensive players and batter.

pitch using a smooth round stick called a bat. After a number of thrown pitches, the batter will either be put out or become a runner on one of the bases. The batter may be put out in several ways:

- He hits a fly ball (a ball in the air) that is caught by one of the fielders.

- He hits a ball in fair territory (explained below) and first base is tagged before the batter reaches first base.

- A third strike (explained below) is caught by the catcher.

Fair territory is the part of the playing field between the line from home plate to first base and the line from home plate to third base. *Foul territory* is the region of the field outside of fair territory. A *strike* is a pitch that is struck at by the batter and missed, or is not struck by the batter and passes through a region called the strike zone. A *ball* is a pitch that is not struck at by the batter and does not enter the strike zone in flight.

A hitter can advance to a runner and reach base safely by:

- Receiving four pitches that are balls. In this case, the batter receives a *walk* or *base-on-balls* and can advance to first base.

- Hitting a ball in fair territory that is not caught by a fielder or thrown to first base before the runner reaches first base. There are different types of hits depending on the advancement of the runner on the play. A *single* is a hit where the runner reaches first base, a *double* is a hit where the runner reaches second base, a *triple* is a hit where a runner reaches third base, and a *home run* is a big hit (usually over the outfield fence) where the runner advances around all bases safely.

One Half-Inning of Baseball

In a half-inning of baseball, the nine players on the offensive team will come to bat in sequence. The players will continue to bat in the inning until three outs are made. To get a flavor of how baseball is played, let us revisit the last game played in the 2001 Major League Baseball season. The New York Yankees and the Arizona Diamondbacks were playing the final game of the World Series-the winner of this game would be declared the best team of the season. We focus on the top half of the 7th inning of the game where the Yankees were batting and the Diamondbacks were leading 1-0. The pitcher for the Diamondbacks was Curt Schilling, who was one of the best pitchers for the National League that season. Figure A.2 displays diagrams of the runner and batter situations for each of the six players that came to bat this particular half-inning.

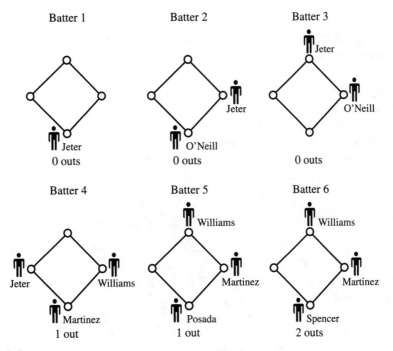

FIGURE A.2
Batter and runner diagrams for each of six New York Yankees players that came to bat in the top of the seventh inning of the final game of the 2001 World Series.

Batter 1: The inning started with no runners on base and no outs. The first batter for the Yankees was Derek Jeter. The first pitch from Schilling to Jeter was a called strike. Jeter hit the second pitch for a single to right field. The Yankees now have a runner on first base with no outs.

Batter 2: The next Yankee batter, Paul O'Neill, hits the first pitch for a single to center field. Jeter advanced to second base. The situation is now runners on first and second with no outs.

Batter 3: Bernie Williams batted next for the Yankees. He hit the first three pitches from Schilling foul (outside of fair territory)—the pitch count now stands at 0 balls and 2 strikes. Williams hit the next pitch to Arizona's first baseman who throws to the second baseman to get O'Neill out; Williams reaches first base on a so-called "fielder's choice." Jeter moves to third base, so the Yankees have runners on first and third with one out and the score is tied at 1-1.

Batter 4: Tino Martinez, the next batter, takes the first pitch from Schilling for a called first strike. Martinez hits the next pitch to right field for a single. Jeter scores from third base and Williams moves to second. Martinez is credited with a "run batted in" as one run scored on the basis of his hit. Now the Yankees have runners on first and second base with one out.

Batter 5: The next hitter, Jorge Posada, takes the first pitch for a called ball. He hits the second pitch for a fly ball to left field. The ball is caught by the outfielder for the second out and the runners stay on first and second.

Batter 6: Shane Spencer, the next batter, hits the first pitch foul for the first strike. He takes the second pitch for the second strike, and hits the third pitch in the air to center field where it is caught by the fielder for the third out.

The Yankees scored one run in their half of the 7th inning to tie the game at 1–1. Arizona eventually won this game 3-2 with an exciting rally in the bottom of the 9th inning and became the baseball champions.

The Boxscore: A Statistical Record of a Baseball Game

One notable aspect of the game of baseball is the wealth of numerical information that is recorded about the game. A boxscore is a statistical record of a particular game. Table A.1 displays a boxscore for the last game in the 2001 World Series. We will define particular baseball events and the associated notation by using this boxscore.

When a player comes to bat during an inning, he is making a *plate appearance* (PA). What can happen during this plate appearance? The batter can get a hit (H) and there are four possible hits: a single (1B), a double (2B), a triple (3B), and a home run (HR). The batter may get a walk (base-on-balls abbreviated BB) by receiving four pitched balls-he advances to first base. Also, the batter can advance to first base when he is hit by a pitch

TABLE A.1

Boxscore of Game 7 of the 2001 World Series between the New York Yankees and the Arizona Diamondbacks.

NY YANKEES (2) VS ARIZONA (3)—FINAL

NY YANKEES	AB	R	H	RBI	BB	SO
Jeter ss	4	1	1	0	0	1
O'Neill rf	3	0	2	0	0	1
[b]Knoblauch ph-lf	1	0	0	0	0	0
B Williams cf	4	0	0	0	0	1
Martinez 1b	4	0	1	1	0	1
Posada c	4	0	0	0	0	2
Spencer lf-rf	3	0	0	0	0	0
Soriano 2b	3	1	1	1	0	1
Brosius 3b	3	0	0	0	0	2
Clemens p	2	0	0	0	0	1
Stanton p	0	0	0	0	0	0
[a]Justice ph	1	0	1	0	0	0
Rivera p	0	0	0	0	0	0
Totals	32	2	6	2	0	10

[a] singled for Stanton in the 8th;

[b] flied to right for O'Neill in the 8th.

BATTING: 2B - O'Neill (1, Schilling).

HR - Soriano (1, 8th inning off Schilling 0 on, 0 out).

ARIZONA	AB	R	H	RBI	BB	SO
Womack ss	5	0	2	1	0	1
Counsell 2b	4	0	1	0	0	0
Gonzalez lf	5	0	1	1	0	2
M Williams 3b	4	0	1	0	0	2
Finley cf	4	1	2	0	0	1
Bautista rf	3	0	1	1	1	1
Grace 1b	4	0	3	0	0	0
Dellucci pr	0	0	0	0	0	0
Miller c	4	0	0	0	0	3
Cummings pr	0	1	0	0	0	0
Schilling p	3	0	0	0	0	3
Batista p	0	0	0	0	0	0
Johnson p	0	0	0	0	0	0
[a]Bell ph	1	1	0	0	0	0
Totals	37	3	11	3	1	13

[a] hit into fielder's choice for Johnson in the 9th.

BATTING: 2B - Bautista (2, Clemens); Womack (3, Rivera).

BASERUNNING: CS - Womack (1, 2nd base by Stanton/Posada).

Ny Yankees	- 000 000 110 – 2
Arizona	- 000 001 002 – 3

One out when winning run scored.

NY YANKEES	IP	H	R	ER	BB	SO	HR	ERA
Clemens	6 1/3	7	1	1	1	10	0	1.35
Stanton	2/3	0	0	0	0	0	0	3.18
Rivera (L, 1-1; BS, 1)	1 1/3	4	2	2	0	3	0	2.84

ARIZONA	IP	H	R	ER	BB	SO	HR	ERA
Schilling	7 1/3	6	2	2	0	9	1	1.69
Batista	1/3	0	0	0	0	0	0	0.00
Johnson (W, 3-0)	1 1/3	0	0	0	0	1	0	1.04

HBP - Counsell (by Rivera). Pitches-strikes: Schilling 103-75; Batista 1-1; Johnson 17-12; Clemens 114-75; Stanton 4-3; Rivera 28-21. Ground balls-fly balls: Schilling 1-11; Batista 1-0; Johnson 1-2; Clemens 6-2; Stanton 0-1; Rivera 1-0. Batters faced: Schilling 27; Batista 1; Johnson 4; Clemens 28; Stanton 1; Rivera 10.

UMPIRES: HP–Steve Rippley. 1B–Mark Hirschbeck. 2B–Dale Scott. 3B–Ed Rapuano. LF–Jim Joyce. RF–Dana Demuth. T–3:20. Att–49,589. Weather: 87 degrees, cloudy. Wind: 18 mph, left to right.

(HBP). The player might create an out. Some outs like a sacrifice bunt (SH) and a sacrifice bunt (SB) advance runners on base. Finally, the player might reach base because of an error by a fielder (E).

An official at-bat (AB) is a plate appearance excluding walks, hit-by-pitches, sacrifice flies, and sacrifice hits. The top half of the boxscore lists all of the batters for both teams. For each player, the boxscore first gives his fielding position. For example, we see that Jeter was the shortstop (ss) for the Yankees in this game. Then the boxscore lists

- AB—the number of at-bats of the player in the game,

- R—the number of runs scored by the player,

- H—the number of hits by the player,

- RBI—the number of runs batted in by the player,

- BB—the number of walks by the player,

- SO—the number of strikeouts by the player,

- LOB—the number of times the player was left on base at the end of the half-inning. If the inning ends with the runner remaining on one of the three bases, then he is said to be left-on-base.

Under the basic batting table, the boxscore lists some special events not included in the table. Under "BATTING", the boxscore lists the players who hit doubles (2B), triples (3B) and home runs (HR). In this listing, the inning in which the hit occurred and the opposing pitcher are recorded. So the listing

2B - O'Neill (1, Schilling)

means that Paul O'Neill hit a double in the first inning against Curt Schilling. Last, the boxscore records special running events. Any successful or unsuccessful stolen bases are listed. (A stolen base is an advancement of a runner to a new base without a play made by the batter.) Here we read

CS - Womack (1, 2nd base by Stanton/Posada).

This means that Tony Womack was unsuccessful in stealing 2nd base and got out—in other words, he was caught stealing (CS). The "1" means that this is the first time he was caught stealing in this series. The Yankees pitcher and catcher were Mike Stanton and Jorge Posada.

After this batting information, the boxscore gives a line summary of the runs and hits scored in the game.

Ny Yankees	- 000 000 110 – 2
Arizona	- 000 001 002 – 3

Each column of numbers corresponds to the number of runs scored by the two teams in a given inning. We see that the Yankees scored a run in each of the 7th and 8th innings, and the Diamondbacks scored one run in the bottom of the 6th and two runs in the bottom of the ninth to win the game.

Below the line score is a table of statistics for the pitchers in the game. For each pitcher, the table gives the number of innings pitched (IP), the number of hits (H) and runs (R) allowed. Next it shows the number of earned runs (ER) allowed—these are runs allowed by the pitcher not due to errors by the fielders of the team. Next the table gives the number of walks (BB) and strikeouts (SO) and home runs (HR) allowed by each pitcher. The last number, the earned run average (ERA), gives the average number of earned runs allowed by this pitcher for nine innings for all games played in the series.

The paragraph below the pitcher table gives some miscellaneous statistics for the pitchers. This paragraph includes a listing of the batters that were hit by a pitch (HBP), the number of pitches, strikes and balls thrown by each pitcher, the number of ground balls and fly balls given up, and the number of batters faced by each pitcher.

Last, the boxscore gives some other information about the game, including the names and positions of the umpires, the elapsed time (T) of the game, the ballpark attendance, and some data on the weather during the game.

B

Datasets Used in the Book and Acquiring Baseball Data over the Internet

Datasets from the Book

Many of the datasets used in the case studies and exercises in Chapters 1-9 can be downloaded from the author's website

http://personal.bgsu.edu/~albert/teachball.htm

The datasets are stored in a text, tab-delimited format. To illustrate, Rickey Henderson's ages and slugging percentages for the first 23 years of his career (the leadoff exercise in Chapter 2) are stored in the file exer_2_0.txt, which looks like the following:

```
AGE     SLG   ←——— Variable names in first row
20      0.336
21      0.399
22      0.437
...
39      0.347
40      0.466      Tab between variable columns
41      0.305
42    ↑ 0.351
```

The file is in text format. Note that the names of all the variables (here AGE and SLG) are contained in the first row and the data follow in successive rows. In each data line, each pair of successive variables is separated by a single tab character. This format is readable by Microsoft Excel and most statistical computing packages such as Minitab and SAS.

Obtaining Baseball Data from the Internet

Ten years ago, it was awkward for the baseball fan to create an electronic file of baseball data. Although books tabulated statistics for players and teams, one had to enter these statistics into a database by hand, and this process was time-consuming and difficult to do without making errors.

Currently, a wealth of baseball data is available over the Internet. All of the sports news sites, such as `espn.com`, `sportsline.com`, `cnnsi.com`, and `sportingnews.com` contain current and historical baseball data. The problem is not to find these data on the Internet, but rather how to download the data into a form suitable for importing into standard statistical computing packages.

Most baseball data in these sports news sites are presented using html tables. This format is helpful for the purposes of displaying the data, but it creates problems when one tries to download the datafile. In the remainder of this section, we discuss the use of three web sites that are especially useful for the person who is interested in downloading baseball data for a particular statistical analysis.

Baseball-reference.com

The web site `http://www.baseball-reference.com`, pictured below, is a very good site for downloading current or historical data on players, teams, and leagues. The data pages in this site are mainly text without any html tables and they are convenient to download and modify to make them readable by statistical computing packages.

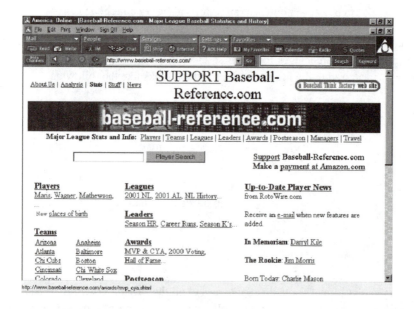

Suppose that you are interested in exploring the season batting averages for Babe Ruth. You do a player search for "Babe Ruth" and you will find a page containing a wealth of hitting and pitching data for Ruth. To download this data, you should

- Save the web page of data in text format.

- Open the page of data using a standard word processor (such as Microsoft Word). Delete much of the textual format of this page, and put this data into a format where

  - The variable names (such as Year and Avg) are on the first line of the file.
  - The data for the variables are stored in separate columns separated by space.

- This data is not in the text, tab delimited format, since there are blanks rather than tabs that separate the data columns. But this data file can be read in a spreadsheet program such as Microsoft Excel. By saving this file in Excel in text format, tabs will be placed between the data columns.

To illustrate the use of the `baseball-reference` site, suppose you are interested in looking at the behavior of run scoring of baseball teams over the years. If we explore the data for "Leagues," we find a table that displays the average number of runs scored per game for the National League for each year of its existence. By saving this table as a text file, importing it into my favorite graphing package (which is MATLAB), I produce the following time series plot of the runs scored per game by the National League teams. There are interesting patterns in this graph. Note that currently (2003), baseball is scoring many runs relative to the last 20 years.

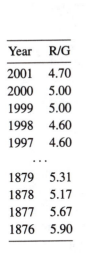

Year	R/G
2001	4.70
2000	5.00
1999	5.00
1998	4.60
1997	4.60
...	
1879	5.31
1878	5.17
1877	5.67
1876	5.90

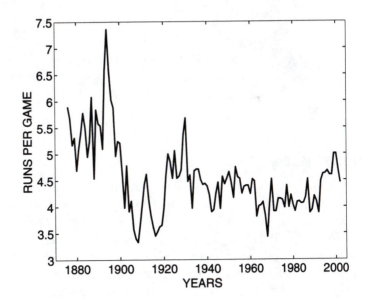

Baseball1.com

The `baseball-reference` web site described above is very useful for obtaining data for individual players, teams in a league, and summary league data across many years. Much of the data in this book is obtainable from this site.

For a longer study that looks at a number of historical players or teams, the Sean Lahman baseball database is a wonderful resource. The complete database is free and can be downloaded from the site *http://www.baseball1.com* shown below. For a nominal cost, the database can be ordered on a cd. This database contains hitting and pitching statistics for *every* player who has played Major League Baseball from 1871–2001. Also, it contains hitting and pitching statistics for every team for each year of its existence. It is currently available in two formats: Microsoft Access and text. The Access database program is convenient for extracting particular parts of the data for analysis.

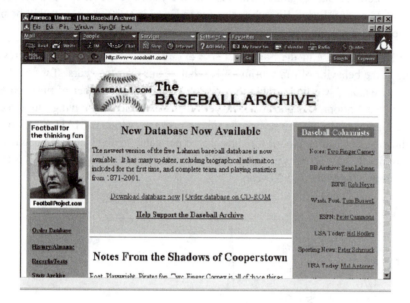

Retrosheet.org

The Lahman database is very useful for obtaining historical season data for players and teams. But this data is summarized by season, and one may be interested in looking at a player's performance during a season or during a game. The Retrosheet organization was founded in 1989 for the purpose of computerizing play-by-play accounts of major league baseball games, and a large quantity of this data is available for free from the `http://www.retrosheet.org` website shown below.

For a particular season, say the 1980 season in the National League, we can download the individual play data. This datafile will contain, for each game played that season, the sequence of batting and non-batting plays for each inning of the game. For each play during the game, the data file will give (among other things)

- the home and away teams, the inning, and the current score,
- the names of the batter and pitcher,
- the runners on base and the number of outs,

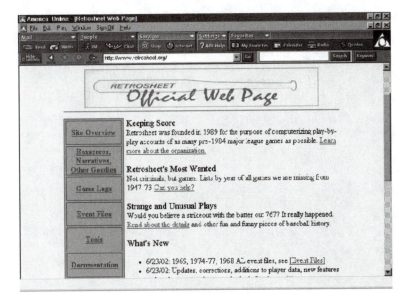

- the names of all of the fielders,

- a description of the play.

The individual play-by-play datafiles are relatively large, and we must run special programs (supplied at the `retrosheet.org` web site), to extract particular variables of interest. These data were used to estimate the transition probabilities in the Markov Chain model of baseball described in Chapter 9.

References

Albert, James. (1998), "Sabermetrics." In *Encyclopedia of Statistical Sciences* (edited by S. Kotz, C. B. Read and D. L. Banks). New York: John Wiley.

Albert, James and Rossman, Allen (2001), *Workshop Statistics: Discovery with Data, A Bayesian Approach.* Emeryville, CA: Key College.

Albert, Jim and Bennett, Jay (2001), *Curve Ball: Baseball, Statistics, and the Role of Chance in the Game.* New York: Copernicus Books.

Bennett, Jay (1998), "Baseball." In *Statistics in Sport* (edited by Jay Bennett). London, New York: Arnold Publishers.

Berry, Donald A. (1996), *Statistics: A Bayesian Perspective.* Belmont, CA: Wadsworth Publishing.

Bukiet, B., Harold, E., and Palacios, J. (1997), A Markov Chain approach to baseball, *Operations Research*, Vol. 45, No. 1, pp. 14–23.

Cover, Thomas M. and Keilers, Carroll. W. (1977), An offensive earned run average for baseball, *Operations Research*, 25, pp. 729–740.

D'Esopo, D. A. and Lefkowitz, B. (1977), "The Distribution of Runs in the Game of Baseball." In *Optimal Strategies in Sports* (edited by S. P. Ladany and R. E. Machol), pp. 55–62. New York: North-Holland.

Devore, Jay and Peck, Roxy (2000), *Statistics: The Exploration and Analysis of Data.* Pacific Grove, CA: Duxbury Press.

Gilovich, T., Vallone, R., and Tversky, A. (1985), The hot hand in basketball: On the misperception of random sequences, *Cognitive Psychology, 17*, 295–314.

James, Bill (1982), *The Bill James Baseball Abstract.* New York: Ballantine Books.

———(1997), *The Bill James Guide to Baseball Managers.* New York: Scribners.

———(2001), *The New Bill James Historical Baseball Abstract.* New York: The Free Press.

Katz, Stanley M., Study of "The Count," *1986 Baseball Research Journal* (#15), pp. 67–72.

Kemeny, J. G. and Snell, J. L. (1976), *Finite Markov Chains*. New York: Springer-Verlag.

Lindsey, G. R. (1963), An investigation of strategies in baseball, *Operations Research*, vol. 11, pp. 477–501.

Major League Handbook 2001. Lincolnwood, IL: Sports Team Analysis & Tracking Systems.

Moore, David and McCabe, George (2003), *Introduction to the Practice of Statistics* (4th edition). New York: W. H. Freeman Company.

Mosteller, F. (1952), The World Series competition, *Journal of the American Statistical Association*, 47, 355–380.

Neft, David S., and Cohen, Richard M. (1997), *The Sports Encyclopedia: Baseball*. New York: St. Martin's Press.

Pankin, Mark D. (1987), "Baseball as a Markov Chain." In *The Great American Baseball Stat Book*, pp. 520–524. (Also see website `http://www.pankin.com/markov/intro.htm`.)

Schell, Michael (1999), *Baseball's All-Time Best Hitters: How Statistics Can Level the Playing Field*. Princeton, NJ: Princeton University Press.

Scheaffer, Richard L. (1995), *Introduction to Probability and its Applications*. Duxbury Press.

Stats Player Profiles 2002. Stats, Inc., Lincolnwood, IL: Sports Team Analysis & Tracking Systems.

Thorn, John and Palmer, Pete (1985), *The Hidden Game of Baseball*. New York: Doubleday.

———— (eds) (1989), *Total Baseball*. New York: Warner Books.

Index

Jim Albert received a BS degree in mathematics from Bucknell University in 1975 and a PhD in statistics from Purdue University in 1979. He has taught at Bowling Green State University since 1979 and is currently a Professor of Mathematics and Statistics. He is a Fellow of the American Statistical Association. His research interests are in Bayesian inference and has published over 60 papers in refereed journals. In addition, he works in the areas of statistical education and applications of statistics in sports. He is currently editor of *The American Statistician,* the "general interest" journal published by The American Statistical Association (ASA). He has been active both in the Section of Bayesian Statistical Science and the Section on Statistics in Sports of the American Statistical Association. He has written four books: *Bayesian Computation Using Minitab, Ordinal Data Modeling* (with Val Johnson), *Workshop Statistics: Discovery with Data, A Bayesian Approach* (with Allan Rossman), and *Curve Ball: Baseball, Statistics, and the Role of Chance in the Game* (with Jay Bennett). The book *Curve Ball* has been reviewed favorably in many publications, including *The Wall Street Journal, The Journal of the American Statistical Association, Physics Today, Technometrics, MAA Online Reviews, Tech Directions Magazine, Baseball American, Mathematics Magazine, SIAM News,* and *Science News. Curve Ball* was a winner of the 2001 The Sporting News-SABR Baseball Research award, and was recently included in the "The Essential Sabermetric Library" in an article by Randy Klipstein in the Newsletter of the Statistical Analysis interest group of SABR (The Society of American Baseball Research).